UNIT

Edexcel AS | 1

Chemistry

The Core Principles of Chemistry

Rod Beavon

Philip Allan Updates, an imprint of Hodder Education, part of Hachette Livre UK, Market Place, Deddington, Oxfordshire OX15 0SE

Orders

Bookpoint Ltd, 130 Milton Park, Abingdon, Oxfordshire OX14 4SB
tel: 01235 827720
fax: 01235 400454
e-mail: uk.orders@bookpoint.co.uk

Lines are open 9.00 a.m.–5.00 p.m., Monday to Saturday, with a 24-hour message answering service. You can also order through the Philip Allan Updates website: www.philipallan.co.uk

ISBN 978-0-340-95012-8

First printed 2008
Impression number 5 4 3 2 1
Year 2013 2012 2011 2010 2009 2008

This guide has been written specifically to support students preparing for the Edexcel AS Chemistry Unit 1 examination. The content has been neither approved nor endorsed by Edexcel and remains the sole responsibility of the author.

Typeset by Fakenham Photosetting Ltd
Printed by MPG Books, Bodmin

Hachette Livre UK's policy is to use papers that are natural, renewable and recyclable products and made from wood grown in sustainable forests. The logging and manufacturing processes are expected to conform to the environmental regulations of the country of origin.

P01296

AS Chemistry

Contents

Introduction

Content Guidance

■ ■ ■

Questions and Answers

Introduction

About this guide

This unit guide is the first of a series covering the Edexcel specification for AS and A2 chemistry. It offers advice for the effective study of **Unit 1: The Core Principles of Chemistry**. Its aim is to help you understand the chemistry — it is not intended as a shopping list, enabling you to cram for an examination. The guide has three sections:

- **Introduction** — this provides guidance on study and revision, together with advice on approaches and techniques to ensure you answer the examination questions in the best way that you can.
- **Content Guidance** — this section is not intended to be a textbook. It offers guidelines on the main features of the content of Unit 1, together with particular advice on making study more productive.
- **Questions and Answers** — this shows you the sort of questions you can expect in the unit test. Answers are provided; in some cases, distinction is made between responses that might have been given by a grade-A candidate and those typical of a grade-C candidate. Careful consideration of these will improve your answers and, much more importantly, will improve your understanding of the chemistry involved.

The effective understanding of chemistry requires time. No-one suggests it is an easy subject, but even those who find it difficult can overcome their problems by the proper investment of time.

To understand the chemistry, you have to make links between the various topics. The subject is coherent; it is not a collection of discrete modules. These links only come with experience, which means time spent thinking about chemistry, working with it and solving chemical problems. Time produces fluency with the ideas. If you have that, together with good technique, the examination will look after itself. Don't be an examination automaton — be a chemist.

The specification

The specification states the chemistry that can be used in the unit tests and describes the format of those tests. This is not necessarily the same as what teachers might choose to teach or what you might choose to learn.

The purpose of this book is to help you with Unit Test 1, but don't forget that what you are doing is learning *chemistry*. The specification can be obtained from Edexcel, either as a printed document or from the web at www.edexcel.org.uk.

The unit test

The unit test lasts 1 hour 15 minutes and is worth 80 marks. This unit counts for 40% of AS or 20% of A2 marks. Section A contains objective test (multiple-choice) questions; Section B contains a mixture of short-answer and extended-answer questions, including questions on the analysis and evaluation of practical work. Quality of written communication is assessed in Section B.

Assessment objectives

Unit Test 1 has three assessment objectives (AOs).

AO1 is 'knowledge and understanding of science and of How Science Works'. This makes up 50% of Unit Test 1. You should be able to:
- recognise, recall and show understanding of scientific knowledge
- select, organise and present information clearly and logically, using specialist vocabulary where appropriate

AO2 is 'application of knowledge and understanding of science and of How Science Works'. This makes up 40% of Unit Test 1. You should be able to:
- analyse and evaluate scientific knowledge and processes
- apply scientific knowledge and processes to unfamiliar situations
- assess the validity, reliability and credibility of scientific information

AO3 is 'How Science Works'. This makes up 10% of Unit Test 1. You should be able to:
- demonstrate and describe ethical, safe and skilful practical techniqes and processes, selecting appropriate qualitative and quantitative methods
- make, record and communicate reliable and valid observations and measurements with appropriate precision and accuracy
- analyse, interpret, explain and evaluate the methodology, results and impact of your own and others' experimental and investigative activities in a variety of ways

How science works (HSW)

HSW is not new. It involves using theories and models, posing questions that can be answered scientifically, carrying out practical investigations and being able to use scientific terminology properly. All this has featured in previous specifications, but in the new one its role in assessment and its proportion in any examination paper are specified. Except for the new material mentioned later, it is unlikely that you would notice any difference between HSW questions and questions on similar topics from past papers.

An example of HSW is in the material on shapes of molecules, in Unit 2. This uses the idea of minimising repulsions between the electron pairs about a given atom. In a molecule, a carbon atom having four electron pairs has a molecular shape that puts these pairs as far apart as possible, so that if all the bonds are single bonds the molecule is tetrahedral, like methane or carbon tetrachloride. This idea can be extended to other molecules to predict their shapes, which is HSW because it is a general theory that enables predictions to be made.

The new material is:
- the assessment of risk in experiments, and understanding the difference between risk and hazard
- considering the way in which science is communicated and how the quality of the work is judged by other scientists ('peer review')
- 'green' chemistry; particularly the efforts to develop processes that are less hazardous and make more efficient use of resources both in terms of the proportion of the atoms that finish up in the product ('atom economy') and in terms of lower energy consumption

Command terms

The following command terms are used in the specification and in unit test questions. You must distinguish between them carefully.
- **Recall** — a simple remembering of facts learned, without any explanation or justification of these facts.
- **Understand** — be able to explain the relationship between facts and underlying chemical principles (understanding enables you to use facts in new situations).
- **Predict** — say what you think will happen on the basis of learned principles.
- **Define** — give a simple definition, without any explanation.
- **Determine** — find out.
- **Show** — relate one set of facts to another set.
- **Interpret** — take data or other types of information and use them to construct chemical theories or principles.
- **Describe** — state the characteristics of a particular material or thing.
- **Explain** — use chemical theories or principles to say why a particular property of a substance or series of substances is as it is.

Learning to learn

Learning is not instinctive — you have to develop suitable techniques to make your use of time effective. In particular, chemistry has peculiar difficulties that need to be understood if your studies are to be effective from the start.

Planning

Busy people do not achieve what they do by approaching life haphazardly. They plan — so that if they are working they mean to be working, and if they are watching television they have planned to do so. Planning is essential. You *must* know what you have to do each day and set aside time to do it. Furthermore, to devote time to study means you may have to give something up that you are already doing. There is no way that you can generate extra hours in the day.

Be realistic in your planning. You cannot work all the time, and you must build in time for recreation and family responsibilities.

Targets

When devising your plan, have a target for each study period. This might be a particular section of the specification, or it might be rearranging information from text into pictures, or drawing a flowchart relating all the reactions of group 1 and group 2. Whatever it is, be determined to master your target material before you leave it.

Reading chemistry textbooks

A page of chemistry may have a range of material that differs widely in difficulty. Therefore, the speed at which the various parts of a page can be read may have to vary, if it is to be understood. In addition, you should read with pencil and paper to hand and jot things down as you go, for example, equations, diagrams and questions to be followed up. If you do not write the questions down, you will forget them; if you do not master detail, you will never become fluent in chemistry.

Text
This is the easiest part to read, and little advice is needed here.

Chemical equations
Equations are used because they are quantitative, concise and internationally understood. Take time over them, copy them and check that they balance. Most of all, try to visualise what is happening as the reaction proceeds. If you can't, make a note to ask someone who can or — even better — ask your teacher to *show* you the reaction if at all possible. Chemical equations describe real processes; they are not abstract algebraic constructs.

Graphs
Graphs give a lot of information, and they must be understood in detail rather than as a general impression. Take time over them. Note what the axes are, what the units are, the shape of the graph and what the shape means in chemical terms.

Tables
Tables are a means of displaying a lot of information. You need to be aware of the table headings and the units of numerical entries. Take time over them. What trends can be seen? How do these relate to chemical properties? Sometimes it can be useful to convert tables of data into graphs.

Diagrams
Diagrams of apparatus should be drawn in section. When you see them, copy them and ask yourself why the apparatus has the features that it has. What is the difference between a distillation and a reflux apparatus, for example? When you do practical work, examine each piece of the apparatus closely so that you know both its form and function.

Mathematical equations

In chemistry, mathematical equations describe the real, physical world. If you do not understand what an equation means, ask someone who does.

Calculations

Do not take calculations on trust — work through them. First, make certain that you understand the problem, and then that you understand each step in the solution. Make clear the units of the physical quantities used and make sure you understand the underlying chemistry. If you have problems, ask.

Always make a note of problems and questions that you need to ask your teacher. Learning is not a contest or a trial. Nobody has ever learnt anything without effort or without running into difficulties from time to time — not even your teachers.

Notes

Most people have notes of some sort. Notes can take many forms: they might be permanent or temporary; they might be lists, diagrams or flowcharts. You have to develop your own styles. For example, notes that are largely words can often be recast into charts or pictures and this is useful for imprinting the material. The more you rework the material, the clearer it will become.

Whatever form your notes take, they must be organised. Notes that are not indexed or filed properly are useless, as are notes written at enormous length and those written so cryptically that they are unintelligible a month later.

Writing

In chemistry, you need to be able to write concisely and accurately. This requires you to marshal your thoughts properly and needs to be practised during your day-to-day learning.

Have your ideas assembled in your head before you start to write. You might imagine them as a list of bullet points. Before you write, have an idea of how you are going to link these points together and also how your answer will end. The space available for an answer is a poor guide to the amount that you have to write — handwriting sizes differ hugely, as does the ability to write succinctly. Filling the space does not necessarily mean you have answered the question. The mark allocation suggests the number of points to be made, not the amount of writing needed.

Re-reading

When you have completed your work, you must re-read it critically. This is remark-ably difficult, because you tend to read what you intended to write rather than what you actually did write. Nevertheless, time spent on the evaluation of your own work is time well spent. You should be able to eliminate most of the silly errors — but you need to practise this in your day-to-day work and not do it for the first time in an examination.

Approaching the unit test

The unit test is designed to allow you to show the examiner what you know. Answering questions successfully is not only a matter of knowing the chemistry, but is also a matter of technique. Unit Test 1 is a paper with multiple-choice, short-answer and extended-answer questions that are answered on the question paper.

Revision

Start your revision in plenty of time. Make a list of the things that you need to do, emphasising the things that you find most difficult, and draw up a detailed revision plan. Work back from the examination date, ideally leaving an entire week free from fresh revision before that date. Be realistic in your revision plan and then add 25% to the timings because everything takes longer than you think.

When revising:
- make a note of difficulties and ask your teacher about them — if you do not make such notes, you will forget to ask
- make use of past papers, but remember that these have been written to a different specification
- revise ideas, rather than forms of words — you are after *understanding*
- remember that scholarship requires time to be spent on the work

When you use the example questions in this book, make a determined effort to answer them before looking up the answers and comments. Remember that the answers given here are not intended as model answers to be learnt parrot-fashion. They are answers designed to illuminate chemical ideas and understanding.

The exam

The exam paper
- *Read the whole paper through before you start to write.* Even though there is no choice of questions, knowing what is around the corner helps the brain to do some very useful subconscious processing.
- Read the question. Questions usually change from one examination to the next. A question that looks the same, at a cursory glance, as one that you have seen before usually has significant differences when read carefully. Needless to say, candidates do not receive credit for writing answers to their own questions.
- Be aware of the number of marks available for a question. That is an excellent pointer to the number of things you need to say.
- Do not repeat the question in your answer. The danger is that you fill up the space available and think that you have answered the question, when in reality some or maybe all of the real points have been ignored.
- The name of a 'rule' is not an explanation for a chemical phenomenon. For example, in equilibrium (Unit 2), a common answer to a question on the effect of

changing pressure on an equilibrium system is 'Because of Le Chatelier's principle...'. That is simply a name for a rule — it does not explain anything.

Multiple-choice questions

Answers to multiple-choice questions are machine-marked. Multiple-choice questions need to be read carefully; it is important not to jump to a conclusion about the answer too quickly. You need to be aware that one of the options might be a 'distracter'. An example of this might be in a question having a numerical answer of, say, $-600\,\text{kJ}\,\text{mol}^{-1}$; a likely distracter would be $+600\,\text{kJ}\,\text{mol}^{-1}$.

Some questions require you to think on paper — there is no demand that multiple-choice questions be carried out in your head. Space is provided on the question paper for rough working. It will not be marked, so do not write anything that matters in this space because no-one will see it.

For each of the questions there are four suggested answers, A, B, C and D. You select the best answer by putting a cross in the box beside the letter of your choice. If you change your mind you should put a line through the box and then indicate your alternative choice. Making more than one choice does not earn any marks. Note that this format is not used in the multiple-choice questions in this book.

The Unit 1 test has 20 multiple-choice questions, which should take you no more than 20 minutes.

Read the paper through

Candidates are, unsurprisingly, eager to get on and write something. Be patient!

Reading the paper through gives the brain an opportunity to do some subconscious processing while the conscious mind is busy directing the writing. It is rather like planning a long walk over the Lakeland fells — you need to have an idea of what is around the corner and to know what dangers to avoid. This technique becomes more important the more chemistry you know, since the A2 papers increasingly expect you to make links between different areas of the specification and to acquire synoptic understanding. These links can be made by the brain in the background during an exam while you are writing other things. Do not underestimate the value of reading the whole paper through *before* you start to write.

Marking

Online marking

It is important that you have some understanding of how examinations are marked, because to some extent it affects how you answer them. Your examination technique partly concerns chemistry and partly must be geared to how the examinations are dealt with physically. You have to pay attention to the layout of what you write. Because all your scripts are scanned and marked online, there are certain things you must do to ensure that all your work is seen and marked. This is covered below.

As the examiner reads your answer, decisions have to be made — is this answer worth the mark or not? Those who think that these decisions are always easy 'because science is right or wrong' have misunderstood the nature of marking and the nature of science.

Your job is to give the *clearest possible answer* to the question asked, in such a way that your chemical understanding is made obvious to the examiner. In particular, you must not expect the examiner to guess what is in your head; you can be judged only by what you write.

Not all marking is the same

The marking of homework is not the same as the marking of examinations. Teachers marking their pupils' work are engaging in formative assessment and their marking is geared towards helping students to improve their understanding of chemistry. It will include comments and suggestions for improving understanding, which are far more important than any mark that might be obtained, as is any discussion resulting from the work. An examination is a summative assessment; candidates have no opportunity to improve, so the mark is everything. That is why questions that are designed to improve the understanding of chemistry during the course are not of the same style as questions used to test that understanding at the end of a course.

Please do not regard examination questions and answers as chemical education in their own right; they are an attempt to see if you have acquired that education through the influence of your teachers and your own reading. An examination is a means to an end, not an end in itself. Education, as distinct from training, is designed to make you (in this case) into a competent chemist rather than one who can simply regurgitate 'model answers' with no underlying understanding. The truth is that if you do the necessary work throughout the course the examination content should look after itself. What you need is the proper answering techniques.

Because examination answers cannot be discussed, you must make your answers as clear as possible. Do not expect examiners to guess what is in your head. This is one reason why you are expected, for example, to show working in calculations. It is especially important that you *think before you write*. You will have a space for your answer on the question paper, and that space is what the examiner has judged to be a reasonable amount of space for the answer. Because of differing handwriting sizes, because of false starts and crossings-out, and because some candidates have a tendency to repeat the question in the answer space, that space is never exactly right for all candidates. Advice on avoiding some of the pitfalls comes later; but the best advice is that *before you begin your answer you must have a clear idea of how it will end*. You do not have time or space for subsequent editing. It is a good plan to practise putting your answer into a list of the points you wish to make, and then join these up into coherent sentences. Or you can leave them as a list; good quality written communication can just as well be presented as a list as a piece of elegant prose.

Common pitfalls

- **Do *not* write in any colour other than black**. This is now an exam board regulation.
- **Do *not* write outside the space provided without saying, *within that space*, where the remainder of the answer can be found**.

Edexcel exam scripts are marked online, so few examiners will handle a real, original script. The process is as follows:

- When the paper is set it is divided up into items, often, but not necessarily, a single part of a question. These items are also called clips.
- The items are set up so that they display on-screen, with check-boxes for the score and various buttons to allow the score to be submitted or for the item to be processed in some other way.
- After you have written your paper it is scanned; from that point all the handling of your paper is electronic. Your answers are tagged with an identity number.
- It is impossible for an examiner to identify a centre or a candidate from any of the information supplied.
- Examiners mark items over a period of about 3 weeks.
- Examiners are instructed on how to apply the marking scheme and are tested to make sure that they know what is required and can mark the paper fairly.
- Examiners are monitored on their performance throughout the marking period. They are prevented from marking a particular item if they do not achieve the necessary standard of accuracy; their defective marking is re-marked.
- Examiners mark items, not whole scripts. Suppose an examiner is marking Q2 (a)(i); depending on the length of this item between 20 and 50 examples will be marked and then the examiner may move on to Q2 (a)(ii). This style of marking means that an individual candidate's paper could be marked by as many as 20 different people.
- Items are allocated to examiners randomly, so generally they do not see more than one item from a given candidate.

The following list of potential pitfalls to avoid is particularly important:

- Do *not* write in any colour other than black. The scans are entirely black-and-white, so any colour used simply comes out black — unless you write in red, in which case it does not come out at all. The scanner cannot see red (or pink or orange) writing. So if, for example, you want to highlight different areas under a graph, or distinguish lines on a graph, you must use a different sort of shading rather than a different colour.
- Do *not* use small writing. Because the answer appears on a screen, the definition is slightly degraded. In particular, small numbers used for powers of 10 can be difficult to see. The original script is always available but it takes a relatively long time to get hold of it.
- Do *not* write in pencil. Faint writing does not scan well.
- Do *not* write outside the space provided without saying, within that space, where the remainder of the answer can be found. Examiners only have access to a given item; they cannot see any other part of your script. So if you carry on your answer elsewhere but do not tell the examiner within the clip that it exists, it will not be seen. Although the examiner cannot mark the out-of-clip work, the paper will be referred to the Principal Examiner responsible for the paper.
- Do *not* use asterisks or arrows as a means of directing examiners where to look

for out-of-clip items. Tell them in words. Candidates use asterisks for all sorts of things and examiners cannot be expected to guess what they mean.

- Do *not* write across the centre-fold of the paper from the left-hand to the right-hand page. A strip about 8 mm wide is lost when the papers are guillotined for scanning.

- Do *not* repeat the question in your answer. If you have a questions such as 'Define the first ionisation energy of calcium', the answer is 'The energy change per mole for the formation of unipositive ions from isolated calcium atoms in the gas phase'; or, using the equation, 'The energy change per mole for $Ca(g) \rightarrow Ca^+(g) + e^-$'. Do not start by writing 'The first ionisation energy for calcium is defined as...' because by then you will have taken up most of the space available for the answer. Examiners know what the question is — they can see it on the paper.

Content
Guidance

This section is a guide to the content of **Unit 1: The Core Principles of Chemistry**.

The main areas of this unit are:
- Formulae, equations and amounts of substance
- Energetics
- Atomic structure and the periodic table
- Bonding
- Introductory organic chemistry

For each part of the specification, you should also consult a standard textbook for more information. Chemistry is a subtle subject, and you need to have a good sense of where the information you are dealing with fits into the larger chemical landscape. This only comes by reading. Remember that the specification tells you only what can be examined in the unit test.

Formulae, equations and amounts of substance

The abilities to calculate formulae from data, write correct formulae, balance equations and perform quantitative calculations are fundamental to the whole study of chemistry. This topic introduces these ideas, and its content could be examined in any unit test.

The ability to lay out calculations in a comprehensible manner depends on:
- knowing what you are doing and not being reliant on rote learning of formulae
- realising that calculations need linking words and phrases to make them readable
- the correct use of units *throughout the calculation*

Using units throughout a calculation is advantageous. It means that you are less likely to get calculations wrong because you are more aware of what you are doing. Some of the advantages are:
- an awareness that equations are not merely symbols, but express relationships between physical quantities
- a check on whether the equations used are in fact correct, because of the in-built check on the units of the answer
- a gradual awareness of what sort of magnitude of answer is reasonable in a given set of circumstances

Some fundamental ideas

- **Atom** — for most purposes the atom is regarded as having a central nucleus containing protons and neutrons. The nucleus is about 10^{-5} of the atomic diameter. The electrons are located in shells of different energy surrounding the nucleus. The shells are made up of orbitals. The important properties of these three particles are given in the table below:

	Mass		Charge	
	Actual/kg	**Relative to proton**	**Actual/C**	**Relative to proton**
Proton	1.673×10^{-27}	1	$+1.602 \times 10^{-9}$	$+1$
Neutron	1.675×10^{-27}	1.001	0	0
Electron	9.11×10^{-31}	1/1836	-1.602×10^{-19}	-1

- **Atomic number** — the number of protons in an atom.
- **Mass number** — the number of protons plus the number of neutrons in an atom. It is always a whole number.
- **Relative atomic mass** — the average mass (i.e. takes into account the abundance of isotopes in the naturally occurring element) of atoms of an element relative to $\frac{1}{12}$ the mass of a carbon-12 atom, which is defined as 12 exactly.

- **Relative isotopic mass** — the mass of an atom of a particular isotope of an element relative to $\frac{1}{12}$ the mass of a carbon-12 atom, which is defined as 12 exactly.
- **Element** — an element is a substance consisting of atoms that all have the same atomic number.
- **Ion** — an ion is an atom or group of bonded atoms that have lost or gained electrons:
  - Loss of electrons means that there are more protons than electrons in the ion, so the charge is positive and the ion is a cation.
  - Gain of electrons means that there are more electrons than protons in the ion, so the charge is negative and the ion is an anion.
- **Molecule** — a molecule consists of two or more atoms covalently bonded in a fixed ratio to form a single entity.
- **Compound** — a compound is a substance consisting of two or more different types of atoms bonded together in a fixed ratio. If the bonding is covalent — as it is, for example, in water, H_2O — then the compound consists of molecules that are single entities containing the stated number of atoms. If the compound is ionic — for example, sodium chloride — the formula shows the simplest ratio in which the atoms combine. In this example, the formula is NaCl. There are no individual NaCl molecules; there is one sodium ion for every chloride ion in the crystal lattice, but no given pair of ions is special.

Amount of substance and the Avogadro constant

In chemistry, the term *amount* has a technical meaning. It refers to the number of moles of substance being considered.

A **mole** is that amount of substance that contains the same number of entities as there are atoms in exactly 0.012 kg (12 g) of the isotope ^{12}C. The entities concerned (molecules, ions, atoms) must also be specified.

The number of entities in a mole of substance is called the **Avogadro constant, N_A.** Its value is $6.02214179 \times 10^{-23}\,\text{mol}^{-1}$.

In 1811, Amedeo Avogadro hypothesised that equal volumes of any gas at the same temperature and pressure contain equal numbers of molecules. Loschmidt (1865) combined his own work with that of Avogadro and calculated the number now known as the Avogadro constant.

The mole is used because chemistry takes place between particles (molecules, ions or atoms). However, the most convenient way of measuring materials in the laboratory is to weigh them. The Avogadro constant, together with the idea of relative molecular mass, provides the necessary link. Water, H_2O, has a relative molecular mass of 18.0. Since 12 g of ^{12}C contains (by definition) N_A carbon atoms, it follows that 18 g of water contains N_A water molecules, since every molecule has a mass 18.0/12 times that of a ^{12}C atom. In fact, for any molecular substance, its relative molecular mass, in grams, contains N_A molecules of that substance. For an ionic substance, its formula mass, in grams, has N_A formula units. This enables reacting masses to be calculated (see page 22).

Molar mass is the mass (in grams) of a substance that contains N_A molecules or, if ionic, N_A formula units, of the substance. It is numerically the same as the relative molecular (formula) mass of the substance.

Concentration

The concentration of a solution can be expressed in several ways, each having its own purpose.

The commonest method is to use **mol dm⁻³**, where the concentration is given as the amount of solute per dm³ of *solution*. When solutions are made up, the required amount of solute is measured out and the solvent is then added until the required volume of solution is obtained. It is not possible to predict the volume change when two materials are mixed to give a solution because it depends on how the particles fit together. For example, mixing 50 cm³ of water and 50 cm³ of ethanol gives only 97 cm³ of aqueous ethanol.

It is possible to calculate the mass of a substance in a known volume of a solution of known concentration, as shown below for 50 cm³ of sodium hydroxide solution of concentration $0.130 \, mol \, dm^{-3}$. Sodium hydroxide has a molar mass of $40 \, g \, mol^{-1}$.

amount of NaOH $= 0.050 \, dm^3 \times 0.130 \, mol \, dm^{-3} = 6.5 \times 10^{-3} \, mol$

mass of NaOH $= 6.5 \times 10^{-3} \, mol \times 40 \, g \, mol^{-1} = 0.26 \, g$

In solutions that are very dilute and in gases such as air containing some airborne pollutant, the amount of material present is small. In such cases the concentration of the solute is given in **parts of solute per million parts of solvent**, by mass. This is abbreviated to **ppm**. A solution having 1 ppm of solute would have $10^{-6}g$ of solute per gram of solution.

Empirical and molecular formulae

The empirical formula of a compound is its formula in its lowest terms. Therefore, CH_2 is the empirical formula of any compound C_nH_{2n}, i.e. alkenes, cycloalkanes and poly(alkenes).

The calculation steps give you:
● the moles of each atom
● the ratio of moles of each atom

To find the **molecular formula**, which is the formula of one molecule of the substance, you need additional information such as the molar mass of the compound.

Example 1

Sodium burns in oxygen to give an oxide containing 58.97% of sodium. What is the empirical formula of this substance?

Answer

Since the compound is an oxide of sodium, there must be 41.03% oxygen in it.

The three steps needed in calculating the empirical formula are given in the table headings.

	Divide by A_r	Divide by smallest	Ratio of atoms
Sodium	58.97/23.0 = 2.564	2.564/2.564 = 1	1
Oxygen	41.03/16.0 = 2.564	2.564/2.564 = 1	1

The compound has an empirical formula of NaO. This is the simplest or lowest ratio of the atoms. The compound is actually sodium peroxide, Na_2O_2 and is not strictly an oxide. In true oxides the oxidation state of the oxygen is -2; in peroxides it is -1.

Example 2

A compound contains 73.47% carbon, 10.20% hydrogen and the remainder is oxygen. The relative molecular mass is 98. Find the empirical and molecular formulae of the compound.

Answer

	Divide by A_r	Divide by smallest	Ratio of atoms
Carbon	73.47/12.0 = 6.1225	6.1225/1.020 = 6	6
Hydrogen	10.20/1.0 = 10.20	10.20/1.020 = 10	10
Oxygen	16.33/16.0 = 1.020	1.020/1.020 = 1	1

The compound has the empirical formula. $C_6H_{10}O$. The mass of this is:

$$(6 \times 12) + (10 \times 1) + 16 = 98$$

So the empirical formula is also the molecular formula.

Example 3

This example shows why you must not do any rounding during the calculation.

A compound contains 39.13% carbon, 52.17% oxygen and 8.70% hydrogen. Calculate the empirical formula.

Answer

	Divide by A_r	Divide by smallest	Ratio of atoms
Carbon	39.13/12.0 = 3.26	3.26/3.26 = 1	1
Oxygen	52.17/16.0 = 3.26	3.26/3.26 = 1	1
Hydrogen	8.70/1.0 = 8.70	8.70/3.26 = 2.66	2.66

The ratios are not whole numbers. This means you have to multiply by a small integer, usually 2 or 3, to get the answer. In this case, multiplying by three gives 3, 3, 8, so the empirical formula is $C_3H_8O_3$.

Chemical equations

The ability to write equations for chemical reactions is important and needs constant practice. Chemical equations, i.e. symbol equations (not word equations which are not equations at all) are central to the language of chemistry. Some reasons for this are that:

- chemical equations are shorter and more pictorial than words
- chemical equations are understood internationally
- chemical equations are quantitative
- chemical equations can give important structural or mechanistic information about compounds or reaction schemes

Whenever you come across a reaction, you should follow these rules:

- learn the equation for it
- check that it balances and make sure that you understand the reason for the presence and amount of each species
- try to visualise what is happening in the reaction vessel

The visualisation of a reaction can be helped by the use of state symbols. The ones commonly used are:

- (s) for solid
- (l) for liquid
- (g) for gas
- (aq) for aqueous solution

Tips When you write state symbols, make them clear. It is common to find exam scripts where the difference between (s) and (g) is small and not always decipherable.

It is not always necessary to use state symbols and can be impossible. Thus the reaction between sodium iodide and concentrated sulfuric acid gives a brown mess that contains several substances in a paste with water and acid. Attempts to use state symbols in such cases are not illuminating. However, in the case of thermochemical equations and in equilibria and redox equations, their use is essential.

Chemical equations are not algebraic constructions — they represent real processes. Do not tack atoms or molecules on here and there to make an equation 'balance'; if it does not describe what really happens, it is wrong.

Ionic equations

Ionic equations have the advantage that they ignore all the species that are not involved in the reaction (i.e. those that are just 'looking on' as the reaction proceeds — hence the name **spectator ions**).

The rules for writing ionic equations are:
(1) Write out the equation for the reaction.
(2) Separate the *soluble* ionic compounds into their ions.
(3) Leave covalent compounds and insoluble ionic compounds as they are.
(4) Delete the common (spectator) ions from both sides of the equation.

This is illustrated by the reaction between solutions of copper sulfate and sodium hydroxide:

$$CuSO_4(aq) + 2NaOH(aq) \rightarrow Cu(OH)_2(s) + Na_2SO_4(aq)$$
$$Cu^{2+}(aq) + \mathbf{SO_4{}^{2-}(aq)} + \mathbf{2Na^+(aq)} + 2OH^-(aq) \rightarrow$$
$$Cu(OH)_2(s) + \mathbf{SO_4{}^{2-}(aq)} + \mathbf{2Na^+(aq)}$$

The spectator ions are in bold. Deleting these gives the ionic equation:

$$Cu^{2+}(aq) + 2OH^-(aq) \rightarrow Cu(OH)_2(s)$$

Reacting masses

Equations can be used to find how much material could be produced from given starting quantities.

Example

Sodium carbonate weighing 5.3 g is reacted with excess hydrochloric acid. Calculate the mass of sodium chloride produced.

Answer

The molar masses of sodium carbonate and of sodium chloride are needed, since the masses of both are involved in the calculation.

molar mass of $Na_2CO_3 = [(2 \times 23.0) + 12.0 + (3 \times 16.0)]\,g\,mol^{-1} = 106.0\,g\,mol^{-1}$

molar mass of $NaCl = 35.5 + 23.0\,g\,mol^{-1} = 58.5\,g\,mol^{-1}$

Then the steps are:

Step 1 Calculate the amount of starting material (sodium carbonate), where amount means moles.

Step 2 Calculate the amount of sodium chloride from the chemical equation.

Step 3 Calculate the mass of sodium chloride.

The equation is:

$$Na_2CO_3 + 2HCl \rightarrow 2NaCl + CO_2 + H_2O$$

$$\text{amount of } Na_2CO_3 = \frac{5.3\,g}{106.0\,g\,mol^{-1}} = 0.050\,mol$$

amount of NaCl $= 2 \times 0.050\,mol = 0.10\,mol$

mass of NaCl $= 0.10\,mol \times 58.5\,g\,mol^{-1} = 5.85\,g$

Most reactions do not give a quantitative yield — there may be competing reactions and there are handling losses unless great care is taken. The **percentage yield** is often calculated and is given by:

$$\% \text{ yield} = \frac{\text{actual yield} \times 100}{\text{calculated yield}}$$

In the example above, an actual yield of 5.21 g sodium chloride would mean a percentage yield of:

$$\frac{5.21\,g \times 100}{5.85\,g} = 89\%$$

Equations from reacting masses

Knowing the mass of the materials that react also gives information about the equation for the reaction. This process was a preoccupation of chemists for much of the eighteenth and nineteenth centuries, since it is the basis on which formulae are determined.

Example

In a suitable organic solvent, tin metal reacts with iodine to give an orange solid and no other product. In such a preparation, 5.9 g of tin reacted completely with 25.4 g of iodine. What is the formula of the compound produced?

Answer

molar mass of tin $= 118.7\,g\,mol^{-1}$

molar mass of iodine = $126.9 \, g \, mol^{-1}$

amount of tin = $\dfrac{5.9 \, g}{118.7 \, g \, mol^{-1}}$ = $0.050 \, mol$

amount of iodine atoms = $\dfrac{25.4 \, g}{126.9 \, g \, mol^{-1}}$ = $0.20 \, mol$

ratio of iodine:tin = 4:1

So, the compound has the empirical formula SnI_4.

The data do not tell us whether the compound is SnI_4, Sn_2I_8 or some other multiple — a molecular mass determination would be necessary to find out. It is actually SnI_4.

The calculation in the above example could also be performed in terms of the numbers of atoms of each substance involved. To do this, the amount of each substance has to be multiplied by the Avogadro constant. The answer is the same since N_A cancels during the calculation. This is always true, which is why we use amount of substance, a small number, rather than the number of atoms, which is an extremely large number.

Calculations involving gas volumes

The molar volume of any gas at a given temperature and pressure is the same. It does not depend on the nature of the gas. At s.t.p., i.e. 0°C and 1atm pressure, the volume is $22.414 \, dm^3 \, mol^{-1}$. There is no loss of principle in taking it to be $24 \, dm^3 \, mol^{-1}$ at 'room temperature and pressure', but don't forget that this is an approximation.

The molar volume of a gas can be used to find the volumes of gases obtained from any particular reaction.

Example 1

This is an extension of the reaction considered previously (page 22).

Sodium carbonate weighing 5.3 g was reacted with excess hydrochloric acid. Calculate the mass of sodium chloride produced and the volume of carbon dioxide obtained at room temperature and pressure.

Answer

The first steps are as before.

$$Na_2CO_3 + 2HCl \rightarrow 2NaCl + CO_2 + H_2O$$

amount of Na_2CO_3 = $\dfrac{5.3 \, g}{106.0 \, g \, mol^{-1}}$ = $0.05 \, mol$

amount of NaCl = 2 × 0.05 mol = 0.10 mol

mass of NaCl = 0.10 mol × 58.5 g mol^{-1} = 5.85 g

The volume of CO_2 produced is given by the amount multiplied by the molar volume:

amount of CO_2 = 0.05 mol

volume of CO_2 = 0.05 mol × 24 dm^3 mol^{-1} = 1.2 dm^3

Example 2

50 cm^3 of propane was burnt in excess oxygen. What volume of oxygen reacts, and what is the volume of carbon dioxide produced?

Answer

In this case, the substances whose volumes are required are gaseous and the initial volume of propane is given. The volumes are in the same ratio as the amount of each substance in the equation:

$C_3H_8(g) + 5O_2(g) \rightarrow 3CO_2(g) + 4H_2O(l)$

molar ratio of the gases propane:oxygen:carbon dioxide = 1:5:3

volume of propane = 50 cm^3

so, volume of carbon dioxide produced = 150 cm^3

It is not necessary to use the molar volume of the gas.

Example 3

27 g of propane was burnt in excess oxygen. What volume of oxygen reacts and what is the volume of carbon dioxide produced?

The molar volume of any gas at the temperature and pressure of the experiment is 24 dm^3 mol^{-1}.

The molar mass of propane is 44.0 g mol^{-1}.

Answer

$C_3H_8(g) + 5O_2(g) \rightarrow 3CO_2(g) + 4H_2O(l)$

amount of propane = $\dfrac{27.0 \text{ g}}{44.0 \text{ g mol}^{-1}}$ = 0.614 mol amount of oxygen = $0.614\,mol \times 5 = 3.07\,mol$

volume of oxygen = $3.07\,mol \times 24\,dm^3\,mol^{-1} = 73.7\,dm^3$

The volume of carbon dioxide produced is calculated thus:

amount of carbon dioxide = $3 \times 0.614\,mol = 1.842\,mol$

volume of carbon dioxide = amount × molar volume

$$= 1.842\,mol \times 24\,dm^3\,mol^{-1}$$

$$= 44.2\,dm^3$$

Percentage yields and atom economies

Until recently, the effectiveness of a synthetic process in chemistry was judged by the percentage yield of the desired compound. Most reactions do not give a quantitative yield — there may be competing reactions, and there will be handling losses unless great care is taken. The percentage yield is given by:

$$\% \text{ yield} = \frac{\text{actual yield of product} \times 100}{\text{calculated yield of product}}$$

With percentage yield, the focus is on the product; the amount of other, possibly useless, materials is not taken into account.

A different measure of the effectiveness of a synthetic process was put forward by Trost (1998). This is the idea of the **atom economy** of a reaction. Atom economy is defined as:

$$\text{atom economy} = \frac{\text{molar mass of the product} \times 100}{\text{molar mass of all the reactants}}$$

The better the atom economy, the greater are the proportions of the starting materials that finish up in the product. The advantages of processes with high atom economies are:
- The consumption (and therefore the cost) of starting materials is minimised.
- There is less production of useless materials in the reaction. This becomes increasingly important as the costs of waste disposal and of compliance with stringent disposal regulations rise rapidly.
- Energy costs are often lower.

Reactions can be categorised according to their atom economies:
- Addition reactions have the highest atom economies since all the reactants are consumed in making the product. Examples include addition polymerisation, addition of HBr to alkenes and the hydrogenation of C=C bonds.
- Substitution reactions have lower atom economies, since one group in a molecule is displaced by another, and the displaced material may be of little use. Examples

include the reaction, in sulfuric acid solution, of sodium bromide with an alcohol to form a bromoalkane. The products include dilute sodium sulfate, which has no commercial value and is discarded. It is not toxic or dangerous — just useless.

- Elimination reactions have the lowest atom economies since they produce at least two materials from one, and one of these is likely to be of little use. An example is the elimination, using KOH in an ethanol solvent, of HBr from a bromoalkane to give an alkene. The products include potassium bromide in dilute aqueous solution — not particularly problematic, but it does have to be disposed of.

Energetics

Enthalpy change

The heat change (ΔH) that occurs during a chemical reaction at constant pressure (i.e. in a vessel open to the atmosphere) is called the **enthalpy change** for that reaction. It is measured in $kJ\,mol^{-1}$, where 'mol^{-1}' refers to the molar quantities given in the equation. Therefore:

$$2SO_2(g) + O_2(g) \rightarrow 2SO_3(g) \quad \Delta H = -92\,kJ\,mol^{-1}$$

means that 92 kJ of heat is given out when 2 moles of sulfur dioxide combine with 1 mole of oxygen to give 2 moles of sulfur trioxide. Such an equation is called a **thermochemical** equation.

Note that an enthalpy change should not be called an energy change. Energy changes during reactions are calculated at constant volume and the values obtained are not necessarily the same as the enthalpy change for the reaction.

Changes of state (e.g. from liquid to gas or from solid to liquid) are accompanied by heat changes. The states of the substances in the reaction are important. For example, the combustion of methane would give out more energy if the product water was liquid rather than gaseous, the difference being the enthalpy of vaporisation of the liquid water. Thermochemical equations are commonly shown for substances in their **standard state**, the associated enthalpy changes being standard enthalpies. This is shown by the superscript symbol '$^{\ominus}$', i.e. $\Delta H^{\ominus}$.

Standard conditions mean:
- all reagents and products are in their thermodynamically most stable state
- 1 atmosphere pressure
- a specified temperature

The value of the temperature is not part of the definition of standard conditions. In practice, if no temperature is shown, the default value is 298 K (25°C).

Some elements have different forms (allotropes) in the solid state. For carbon it is necessary to specify graphite or diamond rather than just solid (s).

Enthalpy level diagrams

Enthalpy level diagrams show the relative energy levels of reactants and products. The horizontal axis is often absent, unlabelled, or may be called the 'reaction coordinate'. Enthalpy level diagrams for exothermic and endothermic reactions are shown below.

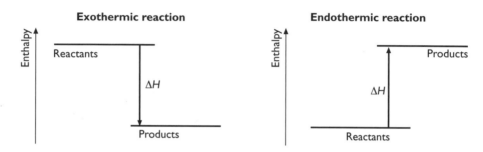

Heat may be given out or be taken in during a chemical reaction. The international convention is that heat changes are seen from the point of view of the chemical system — imagine you are sitting inside the reaction mixture.

- Heat taken in is regarded as positive; so endothermic processes have a positive ΔH.
- Heat given out is regarded as negative; so exothermic processes have a negative ΔH.

Note that the positive or negative signs represent *conventions* concerning the direction of heat flow. They do not represent relative magnitudes in the same way that plus or minus signs do with numbers. Suppose that for one reaction $\Delta H = -100$ kJ mol^{-1} and that for another it is -200 kJ mol^{-1}. The second value is not smaller than the first — indeed it represents twice as much heat given out. Instead of using the terms 'smaller' or 'larger', which are always ambiguous, use descriptive terms, such as 'more exothermic', or whatever is appropriate.

Tips You should always include the sign when you write a value for ΔH, even if it is positive.

Definitions of standard enthalpy changes

- **Standard enthalpy of formation, $\Delta H_f^{\ominus}$**, is the heat change for the formation of 1 mole of substance from its elements, all substances being in their most stable state at 1 atm pressure and a specified temperature.
- **Standard enthalpy of combustion, $\Delta H_c^{\ominus}$**, is the heat change when 1 mole of substance is completely burnt in excess oxygen, all substances being in their most stable state at 1 atm pressure and a specified temperature.

- **Standard enthalpy of neutralisation, $\Delta H_{neut}{}^{\ominus}$**, is the heat change when an acid is neutralised by an alkali, to produce 1 mole of water at 1 atm pressure and a specified temperature.

Strictly speaking, the concentration of the solutions should also be specified when dealing with the standard state, but this refinement is unnecessary for present purposes.

- **Standard enthalpy of atomisation, $\Delta H_{at}{}^{\ominus}$**, is the heat change for the formation of 1 mole of gaseous atoms from an element in its standard state at 1 atm pressure and a specified temperature.

Note carefully that the definition of ΔH_{at} refers to the *formation of 1 mole of gaseous atoms*. Therefore, in the case of chlorine, the heat change measured is for the process:

$$\frac{1}{2} Cl_2(g) \rightarrow Cl(g)$$

Some simple thermochemistry experiments

There are several simple experiments that can be used to measure enthalpy changes. The following account is not intended to substitute for worksheets or other practical instructions. For details you should consult a book of practical chemistry. The principles are as follows:

- Use of a known amount of substance.
- Insulation against heat losses. (Although since the temperature rise or fall for most reactions in solution is fairly small, heat losses are much less than is commonly imagined. The same is not true for combustion reactions (e.g. the burning of alcohols to find the heat of combustion), where the heat losses are colossal.
- Correction of the maximum observed temperature change for heat loss or for slowness of reaction. This is particularly important for reactions between solids and liquids (e.g. heat of displacement in the reaction between copper(II) sulfate solution and zinc metal).
- Calculation of the results.

Determination of enthalpy of neutralisation

Note: This technique can be used for any other type of reaction that can be carried out in an expanded polystyrene cup, so displacement reactions (e.g. between zinc and copper sulfate solution) can also be performed in this way.

The general method for determining the enthalpy of neutralisation is:

- A known volume of acid of known concentration is placed in a polystyrene cup.
- The temperature of the acid is measured for 4 minutes.
- At the fifth minute, a known volume of a solution of a base of suitable concentration is added.
- The temperature is measured every 30 seconds until minute 8.
- A graph of temperature against time is plotted, and the lines are extrapolated to enable the temperature change, $\Delta \theta$, at minute 5, and corrected for any heat losses, to be calculated.

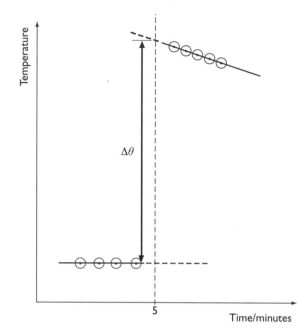

- The heat change (ΔH) is $mc\Delta\theta$, where m is the mass of the solution (usually taken to be equal to its volume for aqueous solutions), c is the heat capacity of the solution (usually taken to be equal to that of water for dilute aqueous solutions), and $\Delta\theta$ is the corrected temperature change.
- The heat change per mole of water formed is then found by dividing the result by the amount (number of moles) of hydrogen ions used.

The method of measuring the temperature change is designed to compensate for two sources of error — cooling (or sometimes warming, since some processes are endothermic and the temperature falls below room temperature) and slowness of reaction. The latter is not a problem in finding enthalpies of neutralisation, since all the species are in solution. However, it can be a problem if the reaction is between a solution and a solid.

Practical details

Using solutions of around $2\,\text{mol}\,\text{dm}^{-3}$ concentration results in a temperature rise of about 14°C, which can be measured with reasonable accuracy.

- Measure $30\,\text{cm}^3$ of $2.00\,\text{mol}\,\text{dm}^{-3}$ hydrochloric acid into an expanded polystyrene cup.
- Measure the temperature of this solution every minute for 4 minutes.
- At the fifth minute, add with stirring $30\,\text{cm}^3$ of $2.20\,\text{mol}\,\text{dm}^{-3}$ sodium hydroxide solution and measure the temperature every 30 seconds from 5.5 minutes to 8 minutes.
- Plot a graph of temperature (y-axis) against time (x-axis).
- Extrapolate the lines on the graph to find the corrected temperature change at 5 minutes.

The heat change can then be calculated using $mc\Delta\theta$.

Enthalpy of combustion of alkanes and alcohols

This simple experiment is not accurate, since the loss of heat to the surroundings is considerable. A set of spirit lamps is used, filled with either a series of liquid alkanes or alcohols.

- A known volume of water is added to a copper calorimeter can.
- The spirit lamp is weighed and placed under the can (without a gauze).
- The temperature of the water is measured and the lamp is lit.
- When the temperature of the water has risen by a suitable amount, the lamp is extinguished and re-weighed.
- The experiment is repeated with another substance in a lamp, using fresh water in the calorimeter.

The heat gained by the water is equal to $mc\Delta\theta$, where m is the mass of water in the calorimeter, c is the heat capacity of water and $\Delta\theta$ is the temperature change. If the mass of the liquid burnt is found from the weighings, then the heat evolved per mole of fuel can be found.

The errors in this experiment are large, arising mainly from the following:

- Heat losses — heat is lost by convection around the calorimeter, so this energy never finds its way into the water.
- Incomplete combustion of the fuel — spirit lamps do not have a good oxygen supply, and the flame, particularly from alkanes, is often sooty.
- Evaporation — some of the fuel evaporates from the spirit lamp, so the mass burnt cannot be determined accurately.

Accurate measurement of combustion enthalpies is carried out using a bomb calorimeter, which is essentially a closed steel vessel containing the fuel with oxygen at high pressure. The calorimeter is completely surrounded by water, so all the heat energy produced on ignition is used to raise the water temperature.

Hess's law

Hess's law states that the enthalpy change for the process Y → Z is independent of the route used to effect the change, provided that the states of Y and Z are the same for each route.

The use of **standard enthalpies** avoids the problem of the states at the beginning and end of the experiment, since the states of Y and Z are defined.

This means that enthalpy changes can be found using other data. For example, enthalpies of reaction can be found from enthalpies of formation or enthalpies of combustion. It is possible, using Hess's law, to find the enthalpy change for reactions that cannot be performed directly.

Note: Hess's law is a particular case of the first law of thermodynamics, which states that matter can be neither created nor destroyed.

Using enthalpies of formation

The Hess's law diagram for calculating enthalpy of reaction from enthalpies of formation is shown below.

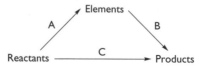

Hess's law says that C = A + B, where A, B and C are the enthalpy changes. Using the definition of enthalpies of formation:

A = – (sum of the enthalpies of formation of the reactants)

B = (sum of the enthalpies of formation of the products)

Therefore:

C = (sum of the enthalpies of formation of the products) – (sum of the enthalpies of formation of the reactants)

Consider the combustion of ethanol:

$$CH_3CH_2OH(l) + 3O_2(g) \rightarrow 2CO_2(g) + 3H_2O(l)$$

The example shows how the heat change for the combustion of ethanol can be found. The states of all substances must be shown, because changes of state also involve enthalpy changes.

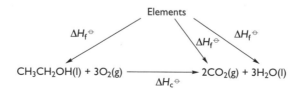

$$\Delta H_c^\ominus = 2\Delta H_f^\ominus (CO_2) + 3\Delta H_f^\ominus (H_2O) - \Delta H_f^\ominus (CH_3CH_2OH)$$

Inserting the appropriate values gives:

$$\Delta H_c^\ominus = 2(-393.5) + 3(-285.8) - (-277.1) = -1367.3\,kJ\,mol^{-1}$$

Note that oxygen is not included, as it is an element in its standard state.

Using enthalpies of combustion

The Hess's law diagram for calculating enthalpy of reaction from enthalpies of combustion is shown below.

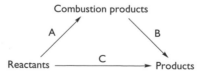

A = (sum of the enthalpies of combustion of the reactants)

B = (sum of the enthalpies of combustion of the products)

Therefore:

C = (sum of the enthalpies of combustion of the reactants) – (sum of the enthalpies of combustion of the products)

Consider the reaction between ethanol and ethanoic acid to give ethyl ethanoate and water:

$$CH_3CH_2OH(l) + CH_3COOH(l) \rightarrow CH_3COOCH_2CH_3(l) + H_2O(l)$$

The Hess's law diagram for this reaction using enthalpies of combustion is shown below:

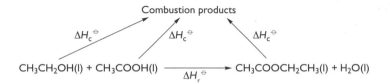

$$\Delta H_r^\ominus = \Delta H_c^\ominus (CH_3CH_2OH(l)) + \Delta H_c^\ominus (CH_3COOH(l)) - \Delta H_c^\ominus (CH_3COOCH_2CH_3(l))$$

Water, of course, does not burn. Inserting the appropriate values gives:

$$\Delta H_r^\ominus = (-1367.3) + (-874.1) - (-2237.9) = -3.5\,kJ\,mol^{-1}$$

Using mean bond enthalpies

Hess's law can be used to find approximate values of $\Delta H^\ominus$ for a reaction by use of mean bond enthalpies. The **mean bond enthalpy** is the enthalpy change when 1 mole of the specified type of bond is broken; an average value is taken which has been determined over a wide variety of molecules. This is the reason why it is approximate — bond enthalpies may vary quite a lot. Thus the C=O bond enthalpy in CO_2 is 805 kJ mol^{-1} but in methanal, HCHO, it is 695 kJ mol^{-1}. Breaking all the bonds in all the reactants leads to a collection of atoms; thus polyatomic elements such as oxygen have to be included in the calculations, unlike in the case of enthalpies of formation.

The Hess's law diagram for the use of bond enthalpies is shown below:

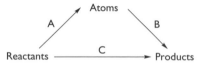

A = (sum of bond enthalpies of the reactants)

B = –(sum of the bond enthalpies of the products)

Therefore:

C = (sum of bond enthalpies of the reactants) – (sum of the bond enthalpies of the products)

Bond enthalpies are always endothermic, i.e. positive.

When working out enthalpies of reaction from bond enthalpies, use the following rules:

• Write the reaction using structural formulae, so that you remember all the bonds.
• Ignore groups that are unchanged between the reactants and the products. In the esterification reaction on page 33, there is no need to involve the CH_3CH_2- group, since it survives unchanged. There is no point in breaking all the bonds only to make them again.

Consider, again, the combustion of ethanol:

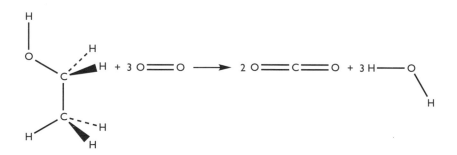

The bonds broken are 5 C–H, 1 C–C, 1 C–O (note the single bond) and 1 O–H. Those made are 4 C=O (note the double bond) and 6 O–H. Using the notation $E(X–X)$ to indicate the average bond enthalpy of the X–X bond and using the equation presented above, together with the bond enthalpies from the *Data Book,* the approximate enthalpy of reaction can be calculated.

approximate enthalpy of reaction

$$= [5E(C–H) + E(C–C) + E(C–O) + E(O–H)] – [4E(C=O) + 6E(O–H)]$$

$$= [(5 \times 413) + 347 + 358 + 464] – [(4 \times 805) + (6 \times 464)]\,kJ\,mol^{-1}$$

$$= -2770\,kJ\,mol^{-1}$$

Experiments on enthalpy changes for reactions that cannot be performed directly

Many Hess's law calculations relate to reactions that cannot be performed directly. For such reactions, knowledge of the enthalpy change, ΔH, is useful in other contexts.

For example, the direct determination of ΔH for the decomposition of sodium hydrogencarbonate on strong heating is not possible. The reaction is:

$$2NaHCO_3(s) \rightarrow Na_2CO_3(s) + CO_2(g) + H_2O(g)$$

An approximate method for finding ΔH is to react sodium hydrogencarbonate and sodium carbonate separately with hydrochloric acid, finding the heat change in each case. The two values obtained can be combined to give an approximate value for ΔH. It is not the actual value because the states of the various reagents are not exactly the same as in the thermal decomposition, but, if desired, adjustments can be made for this using other data, such as hydration enthalpies and enthalpies of solution.

A known amount of sodium hydrogencarbonate is reacted with an excess of hydrochloric acid in an expanded polystyrene cup and the temperature change is determined using methods described previously (page 29). A similar experiment using sodium carbonate is performed and ΔH for each reaction is calculated.

Equation I: $NaHCO_3(aq) + HCl(aq) \rightarrow NaCl(aq) + CO_2(g) + H_2O(l)$ ΔH_1

Equation II: $Na_2CO_3(aq) + 2HCl(aq) \rightarrow 2NaCl(aq) + CO_2(g) + H_2O(l)$ ΔH_2

The overall thermal decomposition reaction approximates to (2 × equation I) – (equation II), so that:

$\Delta H = 2\Delta H_1 - \Delta H_2$

Bond enthalpies and reaction mechanisms

Reaction of halogenoalkanes with hydroxide ions

The rate of reaction between the halogenoalkanes and hydroxide ions in aqueous ethanol depends on the bond strength of the carbon–halogen bond, all other things being equal. If X is a chlorine, bromine or iodine atom:

$CH_3CH_2X + OH^- \rightarrow CH_3CH_2OH + X^-$

The average bond enthalpy of C–Cl is $346\,kJ\,mol^{-1}$, of C–Br is $290\,kJ\,mol^{-1}$, and of C–I is $228\,kJ\,mol^{-1}$.

When chloro-, bromo- and iodoethane are placed separately in solutions of silver nitrate in aqueous ethanol and left in a warm-water bath, the precipitate of silver iodide appears first, followed by that of silver bromide and then, after a much longer time, silver chloride. The stronger the carbon–halogen bond, the slower the reaction, as the activation energy for the reaction increases in the order iodo- < bromo- < chloro-.

Photohalogenation of methane with chlorine

The overall equation for this reaction is:

$CH_4(g) + Cl_2(g) \rightarrow CH_3Cl(g) + HCl(g)$

The initiation step is:

$Cl_2 \rightarrow 2Cl\bullet$

This is followed by one of two possible propagation steps:

Possible step I: $CH_4 + Cl\bullet \rightarrow \bullet CH_3 + HCl$

Possible step II: $CH_4 + Cl\bullet \rightarrow CH_3Cl + H\bullet$

We can use bond enthalpies to predict which propagation step is the more likely:

- Possible step I — a C–H bond is broken ($+435\,kJ\,mol^{-1}$) and an H–Cl bond made ($-432\,kJ\,mol^{-1}$), giving $\Delta H = +3\,kJ\,mol^{-1}$
- Possible step I — a C–H bond is broken ($+435\,kJ\,mol^{-1}$) and a C–Cl bond is made ($-346\,kJ\,mol^{-1}$), giving $\Delta H = +89\,kJ\,mol^{-1}$

Step I, being less endothermic, is the more likely. This is supported by the observed products — step II would give rise to some hydrogen as a result of two H• radicals combining, but none is seen.

Atomic structure and the periodic table

For most purposes the atom is regarded as having a central nucleus containing protons (charge of +1 and mass of 1) and neutrons (charge of zero and mass slightly higher than 1). The diameter of the nucleus is about 10^{-5} of the atomic diameter. The electrons are located in shells of different energy levels surrounding the nucleus. The shells are made up of orbitals.

Definitions

- **Atomic number** — the number of protons in an atom.
- **Mass number** — the number of protons plus the number of neutrons.

> **Tips** The relative isotopic mass is nearly the same as the mass number for a given atom; for carbon-12 it is exactly the same. The difference arises because the neutron and the proton do not have exactly the same mass, and because mass is lost (as energy) when the nucleons come together to form the nucleus. This energy is called the binding energy of the nucleus.

- **Isotopes** – atoms having the same proton number but different mass numbers. They have the same electronic structure and therefore the same chemistry. They differ only in mass.
- **Relative isotopic mass** — the mass of a particular isotope of an element, relative to $\frac{1}{12}$ the mass of a carbon-12 atom, defined as 12 exactly.
- **Relative atomic mass** — the mass of an element relative to $\frac{1}{12}$ the mass of a carbon-12 atom. The relative atomic mass is the weighted mean of the isotopic masses for the element. For chlorine, which is 75% ^{35}Cl and 25% ^{37}Cl, the relative atomic mass is $(0.75 \times 35) + (0.25 \times 37) = 35.5$.
- **Relative molecular mass** — the mass of a molecule relative to $\frac{1}{12}$ the mass of a carbon-12 atom. For an ionic compound, this quantity is sometimes called the relative formula mass.

The mass spectrometer

The mass spectrometer can be used to determine relative atomic mass. It works as follows:
- Gaseous samples of elements or compounds are bombarded with high-energy electrons.
- This causes ionisation, producing positive ions.
- These ions are accelerated and passed through a magnetic field.
- They are bent into a circular path whose radius depends on their mass.
- A detector enables the number of ions at each particular mass to be determined.

The resulting mass spectrum gives information about the isotopic composition of elements or structural information about compounds.

The spectrum of ethanol, CH_3CH_2OH, showing the major fragment ions, is given below:

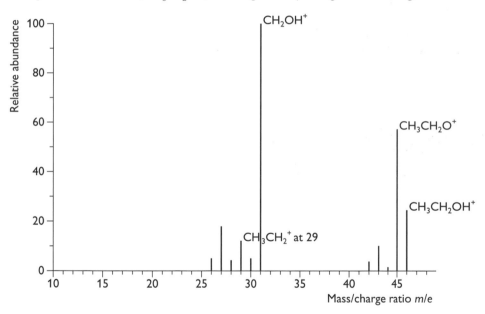

Uses of mass spectral data include:
- determination of the isotopic composition of an element. This was first achieved for neon, when F. W. Aston showed the presence of ^{10}Ne and ^{11}Ne, thus discovering that isotopes exist.
- deduction of the relative atomic mass of an element. This involves finding the masses and abundances of the isotopes, as shown above in the calculation for chlorine above.
- deduction of the structure of an organic compound from the masses of the molecular ions and the fragments that arise from bombardment of the molecules.
- determination of the relative molecular mass of a compound, provided that the molecular ion is present. Not all compounds give a molecular ion peak.

- measurement of the ^{14}C:^{12}C ratio in the radioactive dating of organic material. The older the material, the smaller this ratio is. This application (known as carbon dating) showed that the flax used to make the Turin Shroud, supposedly the burial shroud of Jesus, was grown within a few years of 1305.
- identification of molecules in blood and urine. This application is used on samples taken from athletes for drug testing.

Ionisation energy and electron affinity

Ionisation energy

The **first ionisation energy** for an atom is the energy change involved in the removal of one electron from each of a mole of atoms in the gas phase, i.e. the energy change per mole for:

$$E(g) \rightarrow E^+(g) + e^-$$

The **second ionisation energy** is the energy change per mole for:

$$E^+(g) \rightarrow E^{2+}(g) + e^-$$

Successive ionisation energies are defined similarly.

Successive ionisation energies show the existence of quantum shells; there are significant jumps in ionisation energy between each shell. Thus, for sodium the eleven ionisation energies ($kJ\,mol^{-1}$) are: 496, * 4563, 6913, 9544, 13 352, 16 611, 20 115, 25 491, 28 934, * 141 367, 159 079. The jumps are shown by asterisks, and show one electron in the outer shell, eight in the middle, and two in the innermost.

The first ionisation energies for successive elements in a period are evidence for the existence of sub-shells, where a given shell has electrons of slightly different energy within it. The sub-shells of interest are the s- (2 electrons), p- (8 electrons), and d- (10 electrons) sub-shells. The ionisation energies show discontinuities, for example for the elements in period 3 (Na to Ar) the values are 496, 738 (s-electrons), 578, 789, 1012 (three of the six p-electrons), 1000, 1251, 1521 (the remaining three p-electrons).

Elements of groups 1 and 2 have s-electrons as their outer electrons and constitute the s-block; those of groups 3 to 0 have outer p-electrons and constitute the p-block. In the d-block, the d-sub-shell is being filled — for example, $_{21}Sc$ to $_{30}Zn$ are d-block elements. However, the outer electrons for these elements are s-electrons.

The electron energy levels fill in the following order as atomic number rises as far as krypton (element 36): 1s, 2s, 2p, 3s, 3p, 4s, 3d, 4p. The electrons enter the next sub-shell, once the previous one is full. Thus, $_{23}V$ has the electronic configuration $1s^2$, $2s^2$, $2p^6$, $3s^2$, $3p^6$, $4s^2$, $3d^3$.

Orbitals

Chemistry is concerned with elements bonding together, which involves electrons. The electronic structure therefore determines the chemistry of an element.

The electrons in atoms do not orbit the nucleus like planets around the sun. They exist as volumes of electron density — the **orbitals** — centred on the nucleus. Do not

content guidance

think of orbitals as containers in which an electron bounces around, but rather as a cloud of electron density that can be formed by one electron or by two. The orbitals *are* the electrons.

In orbitals, electrons remain unpaired as far as possible, so, for example, p^3 represents one electron in each of the three p-orbitals.

Covalent bonds are formed by the overlap of orbitals on different atoms, usually one electron each, to give a molecular orbital which can be represented by an electron-density map. This maps the density from the pair of electrons that form the bond. The s-orbital is a sphere centred on the nucleus; each of the three p-orbitals (one along each axis) is rather like two balloons at 180°.

Section through s-orbital   One p-orbital

Electron affinity

The **first electron affinity** of an atom is the energy change for addition of one electron to a mole of atoms in the gas phase:

$$X(g) + e^- \rightarrow X^-(g)$$

This is exothermic, since the electron is attracted to the nuclear charge. The second electron affinity is the energy change per mole for

$$X^-(g) + e^- \rightarrow X^{2-}(g)$$

This is endothermic, since the incoming electron is repelled by the negative charge on the ion.

Periodic properties

The reason that elements in the same group have similar properties is that some properties recur periodically. Ionisation energy is one example, as shown in the graph for period 2 (elements 3–10) and period 3 (elements 11–18) (see page 40). Thus, all the alkali metals have low ionisation energies; the noble (inert) gases have high ionisation energies.

The melting and boiling temperatures of the elements of periods 2 and 3 also show periodic trends. The values for the metals in groups 1–3, with strong metallic bonding, are much higher than those for the non-metals in groups 5–0, but are comparable with those of the giant covalent structures of group 4 elements.

The melting and boiling temperatures of an element reflect the binding energies in the crystal or liquid states. In metals, this depends on how well the outer orbitals of the atoms overlap in the delocalised structure, which depends on the size of the atoms and on the crystal packing. The atoms of metals in period 3 are larger than those in

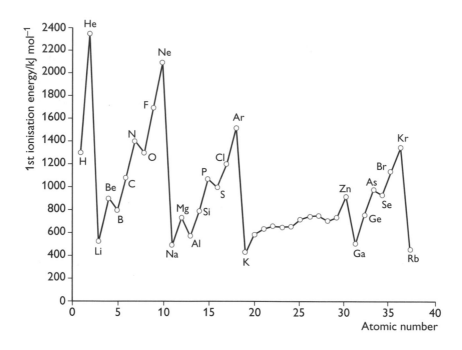

period 2, so the metals have lower melting and boiling temperatures because the atoms are further apart and, therefore, the attractive forces are weaker. With non-metals, the larger the molecules are, the greater the van der Waals forces (covered in Unit 2) and the higher the melting and boiling temperatures.

Periodic trends are shown in the table below.

Element	Li	Be	B	C (graphite)	N	O	F	Ne
Melting temp./°C	181	1278	2300	3697	−210	−218	−220	−248
Boiling temp./°C	1342	2970	2550	4827	−196	−183	−188	−246
Structure	Body-centred cubic	Hexagonal close packing	Tetrahedral	Giant molecule	N_2	O_2	F_2	Atoms

Element	Na	Mg	Al	Si	P	S	Cl	Ar
Melting temp./°C	98	649	660	1410	44	119	−101	−189
Boiling temp./°C	883	1107	2467	2055	280	445	−35	−186
Structure	Body-centred cubic	Hexagonal close packing	Face-centred cubic	Giant molecule	P_4 (white phosphorus)	S_8	Cl_2	Atoms

Bonding

Ionic bonding

The formation of ions

- Positive ions or **cations** are formed by loss of electron(s) from atoms or groups of atoms. The ions are usually metal ions, but can be non-metallic — for example, the ammonium ion, NH_4^+.
- Negative ions or **anions** are formed when atoms or groups of atoms gain electrons.
- The ions formed may have octets of electrons, but many *d*-block metal ions do not. The ion that forms gives the strongest bonding in a resulting compound; the reason that Mg^{2+} is formed rather than Mg^+ is that Mg^{2+} is much smaller than Mg^+, so the cation can get closer to an anion and the attraction is much stronger (see page 45). This increased attraction more than compensates for the extra energy needed to form Mg^{2+} over Mg^+. In many cases, ions *do* have an octet; but such an ion still does *not* have greater 'stability' than its parent atom.
- The driving force for ion formation lies in the lower energy of the compound formed, compared with the constituent elements.

Evidence for the existence of ions

The evidence for the existence of ions in compounds comes from electrolysis. The passage of a current through a molten salt or aqueous salt solution relies on the movement of the ions, which then lose electrons at the anode or gain them at the cathode to form the constituent elements.

This movement can be seen if the ions are coloured. Copper(II) chromate, $CuCrO_4$, contains blue Cu^{2+} and yellow CrO_4^{2-} ions. If a small crystal is placed at the middle of a strip of wetted filter paper attached at each end to a power supply, a yellow colour spreads to the positive electrode and a blue colour to the negative electrode.

Ionic crystals

An ionic crystal consists of a giant three-dimensional lattice of ions. The structure depends on the relative size of the anion and cation, and on the stoichiometry of the substance. Clearly, NaCl (shown on page 42) cannot have the same crystal structure as $MgCl_2$.

The crystal is held together by a *net* attractive force between oppositely charged cations and anions. There are also longer-range repulsive forces between ions of the same charge. However, because these ions are more widely separated the repulsive forces are not as large as the attractive forces between ions of different charge. Because the attractive forces are uniform over the whole crystal:

- ionic substances have high melting and boiling temperatures
- there are no individual molecules in the solid that have the formula of the crystal (e.g. NaCl) — the formula simply shows the relative proportions of the ions in the crystal

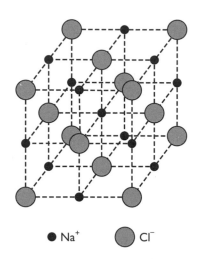

● Na$^+$ ◯ Cl$^-$

The attractive force between two ions is described by Coulomb's law:

$$|F| = \frac{Q_1 Q_2}{k(r_+ + r_-)^2}$$

where $|F|$ is the magnitude of the force between the ions, Q_1 and Q_2 are the charges on the two ions, k is a constant, and $r_+ + r_-$ is the sum of the radii of the two ions and is, therefore, the distance between the centres of the ions when they are touching in the crystal. For a given crystal structure, the attractive force depends on the product of the two charges, and inversely on the square of the distance between the ion centres. This is why a smaller ion, mentioned earlier in connection with Mg^+ and Mg^{2+} (page 41), is important, and why many metal ions form octets of electrons.

Formation of an octet alone does *not* make an ion more stable than the atom from which it arose (see the Born–Haber cycles, page 43).

Ionic radii

Metal ions (cations), formed by electron loss, are smaller than the metal atoms from which they arose. This is because, in forming an ion, the metal atom loses an electron shell.

A non-metal ion formed from a single atom is similar in size to the atom. The addition of extra electron(s) to the outer shell (which already contains electrons) makes little difference to the size. Polyatomic anions, such as sulfate SO_4^{2-}, are much larger.

The size of ions increases going down a group in the periodic table. Successive members of the group have one more shell of electrons. They also have more protons in the nucleus. The combined effect of more shells and more attraction from the nucleus means that size does increase, but not as dramatically as might be expected:

Group 1	Li	Na	K	Rb	Cs
Atomic radius/nm	0.157	0.191	0.235	0.250	0.272
Ionic radius (+1 ion)/nm	0.074	0.102	0.138	0.149	0.150

Group 7	F	Cl	Br	I
Atomic radius/nm	0.155	0.180	0.190	0.195
Ionic radius (−1 ion)/nm	0.133	0.180	0.195	0.215

Isoelectronic ions are ions that have the same electron configuration. Because their nuclei are different, they are different ions. There are six ions that have the electron configuration $1s^2\ 2s^2\ 2p^6$. Their radii (in nm) are shown in the table below:

N^{3-}	O^{2-}	F^-	Na^+	Mg^{2+}	Al^{3+}
0.171	0.140	0.133	0.102	0.072	0.053

The decrease in ionic radius from nitrogen to aluminium is due to the increasing positive charge on the nucleus, which pulls the electron shells closer in.

Born–Haber cycles

The formation of an ionic crystal from its elements can be broken down into a series of stages for which the energy changes can be measured. This enables calculation of the **lattice energy**, which, since it is a measure of the attraction between the ions, is a measure of how strongly the crystal is bonded.

Lattice energy is defined as the energy change for the formation of 1 mole of ionic solid from its isolated gaseous ions.

The lattice energy cannot be measured directly; if an ionic crystal is heated it might vaporise or it might decompose before it does so.

The vapours of ionic materials do not contain separate ions, but rather ion pairs. Thus, gaseous NaCl is Na^+Cl^-, and the energy needed to form ion pairs is not the same as the lattice energy.

However, there are other processes for which the energy change can be measured and which, when put together in a Born–Haber cycle, enable a value for the lattice energy to be calculated. Such values are called **experimental values**. The lattice energy can also be calculated from the geometry of the crystal and the use of Coulomb's law, which gives a value for the **theoretical lattice energy**, or the lattice energy for a *purely* ionic model. Actual lattice energy and theoretical lattice energy are seldom the same, because most ionic crystals also have some covalent bonding.

The Born–Haber cycle for sodium chloride is shown below:

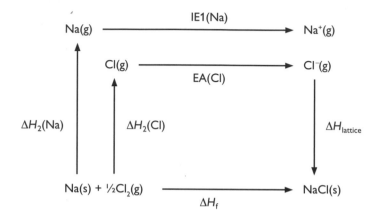

Polarisation of ions: covalence in ionic lattices

Ionic and covalent bonds are extremes. Ionic bonds involve complete electron transfer and covalent bonds involve equal sharing of electrons. In practice, bonding in most compounds is intermediate between these two forms, with one type being predominant.

The attraction that a bonded atom has for electrons is called its **electronegativity**. There are several electronegativity scales, the commonest being the Pauling scale. Fluorine is the most electronegative element with a value of 4, caesium the least at 0.7.

- Atoms with the same electronegativity bond covalently, with equal sharing of electrons.
- Atoms with different electronegativities form polar covalent bonds if the difference is not too large — up to about 1.5 — and ionic bonds if the difference is more than about 2.

Cations are able to distort the electron clouds of anions, which are generally larger. This leads to a degree of electron sharing and hence some covalence in most ionic compounds. Small cations with a high charge (2+ or 3+) have a high charge density. They can polarise anions, and therefore have a high **polarising power**. Large anions do not hold on to their outer electrons very tightly, so they can be easily distorted — such ions are **polarisable**.

The effect of polarisation is that the lattice energy of these compounds is different from that predicted on the basis of a purely ionic model; the lattice energy is more exothermic, so the lattice is stronger. The smaller the cation associated with a given anion, the greater is the degree of covalency. This is because smaller cations have higher charge densities.

Compound	$\Delta H_{latt}/kJ\,mol^{-1}$ (ionic model)	$\Delta H_{latt}/kJ\,mol^{-1}$ (Born–Haber)	Difference/$kJ\,mol^{-1}$
$MgCl_2$	−2326	−2526	200
$CaCl_2$	−2223	−2258	35
$SrCl_2$	−2127	−2156	29
$BaCl_2$	−2056	−2033	23

The larger the anion associated with a given cation, the greater is the degree of covalency. This is because the larger anions are more polarisable.

Compound	$\Delta H_{latt}/kJ\,mol^{-1}$ (ionic model)	$\Delta H_{latt}/kJ\,mol^{-1}$ (Born–Haber)	Difference/$kJ\,mol^{-1}$
MgF_2	−2913	−2957	44
$MgCl_2$	−2326	−2526	200
$MgBr_2$	−2097	−2440	343
MgI_2	−1944	−2327	383

Why $MgCl_2$ rather than MgCl or $MgCl_3$?

The crystal formed from magnesium and chlorine is that which maximises the lattice energy. It has the formula $MgCl_2$.

It is true that less energy is needed to form Mg^+ than is needed to form Mg^{2+}. However, the 2+ ion has lost an electron shell and is half the size of the Mg^+ ion. Thus, the 2+ cation can approach the chloride anion more closely. It also has double the charge of Mg^+. Taken together, these two factors make the lattice energy increasingly exothermic. The attraction of Mg^{2+} for the chloride anion is, therefore, much greater than that of Mg^+ and so the lattice energy for $MgCl_2$ ($-2526\,kJ\,mol^{-1}$) is much more exothermic than that for MgCl (estimated at about $-700\,kJ\,mol^{-1}$).

$MgCl_3$ would require the formation of Mg^{3+}, which is $7733\,kJ\,mol^{-1}$ more endothermic than the formation of Mg^{2+}, but with hardly any change in size. The attraction of Mg^{3+} for the chloride ion is not sufficiently greater than the attraction of Mg^{2+} for the extra ionisation energy to be recovered by an increase in lattice energy.

Covalent bonding

Covalent bonds are strong and have a number of potential features:
- Electrons are shared in pairs.
- A bond is formed because of the attraction of the bonding electrons to the nuclei of both bonded atoms.
- Two or three electron pairs can be shared between a pair of atoms to give a double or triple bond.

- The bonded atoms often contain octets but some have only six electrons in the valence (outer) shell — for example, in BCl_3 there are only six electrons round the boron atom. Atoms in period 3 and beyond can have more than eight electrons since d-orbitals are available — for example, S in SF_6 has 12 electrons in the valence shell.
- Dative covalent bonds are formed when both bonding electrons come from the same atom.
- Dative bonds are no different from any other covalent bond.
- Dative bonds are found in NH_4^+; between chlorine and electron-deficient aluminium in Al_2Cl_6; and as the bonds between water ligands and the central metal ion in hexaqua ions of metals such as $[Mg(H_2O)_6]^{2+}$ and $[Fe(H_2O)_6]^{2+}$.
- Electrons do not 'circle around' nuclei — they exist as charge clouds or orbitals. Covalent bonds consist of overlapping orbitals. Ordinary covalent bonds are overlapping one-electron orbitals, whereas dative bonds are from overlap of a two-electron orbital with an 'empty' orbital.

Giant atomic structures such as diamond and graphite illustrate the strength of covalent bonds:

- All the bonds in the diamond crystal are covalent. It has layers of hexagonal rings that are puckered (i.e. not flat); each carbon atom is bonded covalently to four others throughout the lattice. A great deal of energy is needed to break these bonds, so diamond has an extremely high melting temperature (about 3800°C). There are no free electrons, so diamond is a poor electrical conductor. It is a good thermal conductor, since the rigid lattice readily transmits vibrations.
- Graphite has layers of flat hexagons with a fourth bond delocalised along the plane of the carbons. The forces between the layers are van der Waals forces and are, therefore, weak. Graphite is really a stack of giant molecules, rather like a pile of paper. It is a good electrical conductor parallel to the planes of carbon atoms, but a poor thermal conductor. It sublimes at 3730°C.

Covalent compounds that are small molecules have strong bonds *between atoms* — intramolecular bonds — but much weaker forces *between molecules*. They have, therefore, low melting and boiling temperatures because little energy is required to overcome the **intermolecular forces** between molecules.

Electron density maps

The electron density map for a molecule shows the electron density per unit volume as a set of contours. The map for hydrogen is shown below, with electron density increasing the nearer the contour is to the nucleus. The covalent bond forms because the electron density envelops both atoms.

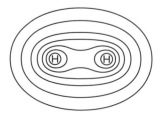

In ionic compounds, the electron densities from the cation and the anion are separate, as in sodium chloride. The chloride ions are larger than the sodium ions. The electron density map for sodium chloride is shown below:

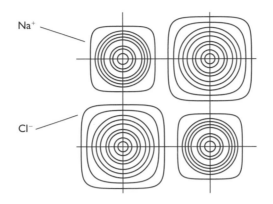

Dot-and-cross diagrams

The electron structure of a covalent molecule can be drawn as a dot-and-cross diagram (Lewis structure), in which the electron pairs are shown as dots and crosses. The electrons are of course identical, so a diagram with all dots or all crosses would also be a correct representation. Some examples are shown below.

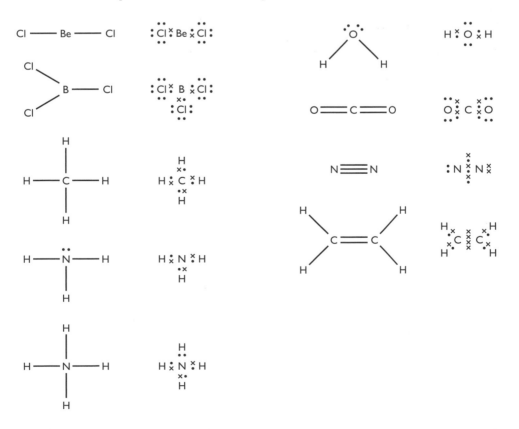

Metallic bonding

The Drude–Lorentz model of a metallic crystal shows the following features:
- Metal ions are usually held in a close-packed lattice, the exception being group 1 metals, which are not close-packed.
- The outer (valence) electrons are delocalised throughout the metal crystal.
- The bonding is not particularly directional. This means the lattice can be distorted without breaking, so the metal is malleable and ductile.
- Mobile electrons enable conduction of electricity in the solid. However, the model does not account for the widely differing conductivities of metals.
- The more delocalised electrons there are, the stronger the bonding and the higher the melting temperature.
- The smaller the atom, the closer the packing and the higher the melting temperature are.

Introductory organic chemistry

Introduction

Organic chemistry is dominated by a number of **homologous series**, arising from the presence of one or more functional groups (pages 49–50) in a molecule. The huge variety of organic compounds arises from the fact that organic chemistry is rather like atomic Lego, as you will realise if you've used molecular model kits.

Homologous series

A homologous series is a series of compounds that:
- have a common general formula
- differ by CH_2
- show a trend in physical properties, for example boiling temperature
- show similar chemical properties since all members of the same homologous series have the same functional group

The first member of a homologous series often shows differences in the detail of its chemistry from successive members of the series.

The **alkenes** are an example of a homologous series in that:
- their general formula is C_nH_{2n} (though not all compounds with this general formula are alkenes)
- they have a C=C double bond (which other compounds with the same general formula do not have)
- there is no alkene with only one carbon atom since there could not be a C=C bond

The first three straight-chain compounds in the series are ethene, $CH_2=CH_2$, propene, $CH_3CH=CH_2$, and butene. Butene has structural and geometric isomers (see below), one of which is but-1-ene, $CH_2=CHCH_2CH_3$. The boiling temperatures of these three alkenes are $-104°C$, $-47.7°C$ and $-6.2°C$ respectively. All three react rapidly with bromine water, with potassium manganate(VII) solution, and with all the other reagents given on pages 64–67.

Functional groups

A **functional group** is a small group of atoms or perhaps a single atom that determines the chemistry of a molecule. The homologous series and functional groups appropriate to AS and A2 are given in the table below. R represents an organic group.

Series	General formula	First member
Alkanes	C_nH_{2n+2}	CH_4, methane
Alkenes	C_nH_{2n}	$CH_2=CH_2$, ethene
Halogenoalkanes	R**X** where **X** is –Cl, –Br or –I.	CH_3Cl, chloromethane (or bromo- or iodo-)
Primary alcohols	RCH_2OH	CH_3OH, methanol
Secondary alcohols	where R and R' may be the same	$CH_3CH(OH)CH_3$, propan-2-ol
Tertiary alcohols	where the Rs may be the same	$(CH_3)_3COH$, 2-methylpropan-2-ol
Aldehydes	**RCHO**	HCHO, methanal
Ketones	**RCOR'** R and R' can be the same	CH_3COCH_3, propanone
Carboxylic acids	**RCOOH**	HCOOH, methanoic acid
Acid chlorides	**RCOCl**	CH_3COCl, ethanoyl chloride

Series	General formula	First member
Esters	R**COOR'** R\C=O / R'O R and R' can be the same. R, but not R', can be H.	$HCOOCH_3$, methyl methanoate
Primary amines	R**NH$_2$**	CH_3NH_2, aminomethane
Nitriles	R**CN**	CH_3CN, ethanonitrile
Amides	R**CONH$_2$** R\C=O / H$_2$N	CH_3CONH_2, ethanamide
Amino acids	NH$_3^+$ R—C—COO$^-$ R' The Rs may be the same or different or may be H, as in glycine	$^-OOCCH_2NH_3^+$, glycine. This form is called a zwitterion, where the acidic COOH group has given its proton to the basic NH_2 group.

Naming and drawing organic compounds

In organic chemistry, the structure of a compound is central to its chemistry. All but rather simple molecular formulae give rise to a variety, sometimes enormous, of different structures.

Nomenclature

The International Union of Pure and Applied Chemistry (IUPAC) has generated a series of rules giving organic compounds **systematic names**. Knowing the rules and the systematic name of a compound means the formula can be written, and vice versa.

The names may be long if the compounds are large. Systematic names are used in the Edexcel specification for most organic substances, though there are exceptions, for example the amino acid, glycine. However systematic names are less common in chemistry beyond school level. This is partly because American textbooks and chemical suppliers do not often use them and partly because the names are unwieldy if more than five or six carbon atoms are involved.

The rules for naming alkanes are given below, with some examples. These rules are built on to name other types of organic molecule.

- Identify the longest carbon chain — take care with this since structures can be written with 90° bond angles that may mask the longest chain. Thus the compound

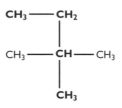

has a four-carbon longest chain, not three carbons.
- The name is based on the alkane with the same number of carbon atoms as the longest chain.
- Substituent groups have the name ending changed to -yl.

This is illustrated for alkanes up to C_5 in the table below.

C	Alkane	Name stem	Substituent group	Structure
1	Methane, CH_4	Meth-	Methyl	CH_3-
2	Ethane, C_2H_6	Eth-	Ethyl	CH_3CH_2-
3	Propane, C_3H_8	Prop-	Propyl	$CH_3CH_2CH_2-$
4	Butane, C_4H_{10}	But-	Butyl	$CH_3CH_2CH_2CH_2-$
5	Pentane, C_5H_{12}	Pent-	Pentyl	$CH_3CH_2CH_2CH_2CH_2-$

- The position of each substituent group is indicated by a number. This corresponds to the position in the chain of the carbon atom to which it is attached, so that the number of the substituent group is the lowest possible. However, for substances such as carboxylic acids, RCOOH, where the functional group is at the end of a chain, the carbon atoms are always numbered from that end.
- Different series of compounds have the name-ending modified to indicate which homologous series is involved.
- Where two groups that are the same are substituted on a given carbon atom, the number is repeated and the substituent is prefixed 'di'.
- Where two groups that are the same are on different carbon atoms, 'di' is still used; if there are three groups the same on different carbons then 'tri' is used; if there are four groups then 'tetra' is used, and so on.

Examples of systematic names

Alkanes

CH₃
|
CH₃CH₂CHCH₃

2-methylbutane

CH₃ CH₃
| |
CH₃CCH₂CHCH₃
|
CH₃

2,2,4-trimethylpentane

Alkenes

The position of the double bond (or bonds) is given by the number of the carbon atom that the double bond starts from.

$CH_3CH=CH_2$
Propene

$CH_3CH_2CH=CH_2$
But-1-ene

$CH_3CH=CHCH_3$
But-2-ene

$CH_2=CH-CH=CH_3$
Buta-1,3-diene; not an alkene, but an alkadiene

Halogenoalkanes

$CH_3CH_2CH_2Cl$
1-chloropropane

$BrCH_2CH_2Br$
1,2-dibromoethane

CH_3CHCH_3
|
Br
2-bromopropane

Alcohols

$CH_3CH_2CH_2OH$
Propan-1-ol

$HOCH_2CH_2OH$
Ethane-1,2-diol

CH_3CHCH_3
|
OH
Propan-2-ol

Aldehydes

$CH_3CH_2CH_2CHO$
Butanal

CH_3
|
$CH_3CHCH_2CH_2CHO$
4-methylpentanal

Ketones

CH_3COCH_3
Propanone

$CH_3COCH_2CH_2CH_3$
Pentan-2-one

$CH_3CH_2COCH_2CH_3$
Pentan-3-one

Carboxylic acids

$CH_3CH_2CH_2COOH$
Butanoic acid

$CH_2=CHCOOH$
Propenoic acid

$CH_3CH=CH_2COOH$
But-2-enoic acid

CH_3CH_2COOH
Propanoic acid — note the difference

CH_3CHCH_2COOH
|
OH
3-hydroxybutanoic acid

$HOOC-COOH$
Ethanedioic acid

Drawing organic molecules

The molecular formula of an organic compound may not be particularly useful, since there is usually more than one possible structure — for example, C_4H_{10} has three possible structures. This problem is overcome by presenting formulae in different ways:

- **Structural formulae** show most of the structure, but not all of the bonds are shown. The side-chain substituents are shown in parentheses.
- **Displayed formulae** show all of the bonds.
- **Skeletal formulae** are used widely for natural products and other large molecules. They show the carbon skeleton by lines (which represent the bonds joining the atoms), but do not show the carbon and hydrogen atoms. All other types of atom are shown.

Consider the molecule 2-methylbutane. Four representations of its structure are shown in the diagram below:

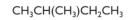

$$CH_3CH(CH_3)CH_2CH_3$$

Structural formula

Semi-displayed formula

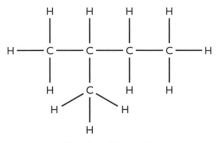

Displayed formula

Skeletal formula

Tips The structural formula with the side-chain substituents in parentheses is useful for typing on a single line. The semi-displayed formula would be used in hand-written material.

The example of *cis*-but-2-enoic acid illustrates how to represent C=C double bonds and atoms other than carbon:

$$CH_3CH=CHCOOH$$

Structural

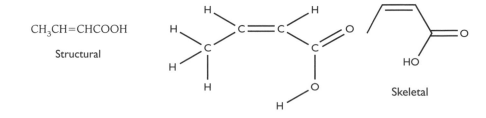

Skeletal

Displayed

Safety in practical chemistry: hazard and risk

Safety is important in chemistry for obvious reasons. Chemistry nevertheless deals with compounds that may be hazardous, so that part of a chemist's training is in the safe use of hazardous materials. **Hazard** is not the same as **risk** — to take a simple example of the difference, a deep lake with treacherous currents is a hazard, but it poses no risk unless you happen to be on it or in it. Hazard is an intrinsic property; risk is personal.

In any question on safety, you would be expected to suggest safety precautions specific to the experiment under consideration, and to be able to assess which is the most significant of various hazards presented.

In this context, the use of lab coats, safety glasses and pipette fillers is considered routine good practice and does not receive credit in examination answers.

The following is presented so as to give you the idea of the basis on which risk assessments might be made. *Safety will not be examined in this detail*. However, comments made by students on safety matters tend to be either so trivial as to be not worth reading, or so apocalyptic as to suggest that all chemistry should cease forthwith.

Whether a procedure is safe or not depends on:
- the intrinsic hazards presented by the chemicals used
- the scale on which the experiment is to be performed
- the containment regime used — for example, whether the experiment is to be done in an open laboratory or in a fume cupboard
- the time of exposure, though this does not affect the initial estimate of risk
- the skill and experience of the experimenter

Every experiment should be the subject of a risk assessment by the institution that intends to perform it. All the above contribute to the final assessment as to whether the risk presented is acceptable, and whether the experiment is therefore feasible. The following is included to illustrate one particular approach, suggested by the Royal Society of Chemistry for use in schools.

Hazard evaluation

The **hazard** associated with a substance is its potential to impair health. Some degree of hazard can be ascribed to almost any substance, while for some the toxicity or the harmful effects are not known fully.

Substances that are likely to be hazardous are those that are
- very toxic (including carcinogenic materials)
- toxic
- harmful
- corrosive
- irritant

Hazardous substances also include those that:
- have a maximum exposure limit (MEL)
- have an occupational exposure standard (OES)

- may produce dusts in appreciable concentration (typically $10\,\mathrm{mg\,m^{-3}}$ total inhalable dust).

Hazard categories are used to help the hazard evaluation:

Hazard category	Hazard classification
Extreme	(None kept in schools)
High	Very toxic; toxic; defined MEL or OES; substances with unknown toxicity
Medium	Harmful; irritant; corrosive
Low	Substances not meeting criteria for hazard labelling

Each substance to be used should be checked and its hazard category noted on the risk assessment sheet for the proposed experiment.

Exposure potential

Hazardous substances vary enormously in potency. A substance with a high hazard may therefore present an acceptably small risk if the exposure potential is low. Conversely, unacceptable risks may result from high exposures to substances with low hazard.

Factors to be taken into account in evaluating exposure potential relate to both **substance** and **activity**.

Substance factors include:
- quantity used
- physical form and properties
- volatility
- dustiness
- concentration if in solution

Activity factors include:
- potential for exposure (e.g. production of aerosol)
- route of exposure (skin, inhalation, ingestion)
- frequency and duration of activity

Small-scale working is preferred wherever possible.

content guidance

Typical basis for estimating exposure potential

The normal basis for estimating exposure potential is shown in the matrix below.

Score	1	10	100
(A) Quantity of substance	Less than 1 g	1–100 g	More than 100 g
(B) Physical character of substance	Dense solids; non-volatile liquids; no skin absorption	Dusty or lyophilised solids; volatile liquids; low skin absorption	Gases; highly volatile liquids; aerosols; solutions that promote skin absorption
(C) Characteristics of operation or activity	Predominantly enclosed system; low chance of mishap	Partially open system; low chance of mishap	No physical barrier; any operation where chance of mishap is medium or high

The exposure potential is estimated by multiplying A × B × C:

ABC < 1000 = *low* exposure potential

1000 < ABC < 10 000 = *medium* exposure potential

10 000 < ABC = *high* exposure potential

Time factors such as the frequency and duration of an activity should be considered. Short-duration tasks involving a few seconds exposure at infrequent intervals should not affect the initial estimate. Continuous operations on a daily basis would probably raise the estimate to the next higher category.

Risk assessment

The **risk assessment** is made using the matrix below, where risk = hazard × exposure potential.

	Exposure potential		
Hazard category	**Low**	**Medium**	**High**
High	2	2/3	2/3
Medium	1	2	2
Low	1	1	1

There is also an *'extreme* hazard' category. Risks presented by substances in this category are unsuited to this evaluation procedure and must be addressed on an individual basis. (They would automatically be unsuitable for use in schools.)

A containment regime is then devised:

- Experiments with a risk assessment of 1 can be carried out on an open bench.
- Those with a risk assessment of 2 require a fume cupboard or other specially vented area.
- Those with a risk assessment of 3 need a special facility. This would not usually be available in schools, so experiments falling in this category could not be carried out.

Either the appropriate containment regime is selected, or the experiment is modified until its risk is acceptable.

Perhaps one misunderstanding could be dispelled here. Benzene is carcinogenic and its use is not allowed in schools. This does *not* mean that aromatic compounds in general are carcinogenic — some are, but far more are not. Many natural products and metabolic intermediates are aromatic compounds containing the benzene ring.

Flammable substances

You may have noticed that the above deals with toxicity only. There is another risk that may be more significant for short-term, small-scale experiments — flammability. Flammable substances must be kept away from naked flames, so any heating necessary must be with a hot-water bath (not heated from underneath by a Bunsen burner) or an electric heating mantle.

Common hazards in school chemistry

The following list is by no means exhaustive or definitive.

Halogens are toxic and harmful by inhalation, although iodine is less toxic than chlorine or bromine, because it is a solid. Chlorine and bromine must always be used in a fume cupboard. Liquid bromine causes serious ulcerating burns and must be handled with gloves, so is best left to demonstration experiments by the teacher.

Ammonia is toxic. Concentrated ammonia solutions should be handled in the fume cupboard.

Concentrated mineral acids are corrosive. If spilt on the hands, washing with plenty of water — *ncvcr* alkali, which is even more damaging — is usually enough, but advice must be sought. Acid in the eye requires *immediate* and copious irrigation and immediate professional medical attention.

Barium chloride solution is extremely poisonous, as are **chromates** and **dichromates**.

Sodium hydroxide, potassium hydroxide or **concentrated ammonia** in the eye is *extremely serious*, and must always receive professional and *immediate* medical attention following copious irrigation of the eye. Sodium hydroxide and other alkali metal hydroxides are among the most damaging of all common substances to skin and other tissue. Treat them with respect.

Phenol is damaging to skin and should always be handled with gloves.

School chemistry laboratories possess data sheets on the hazards associated with chemicals. They may have CLEAPSS hazard cards produced specifically for schools; otherwise, every chemical supplier sends hazard data with any chemical that has been ordered.

None of the above is intended to prevent people doing chemistry — it is intended to encourage safe working practices.

In summary, risk can be minimised by:
- working on a smaller scale
- taking specific precautions or using alternative techniques depending on the properties of the substances involved
- carrying out the reaction using an alternative method that involves less hazardous substances

Alkanes

- **Alkanes** (also known as the paraffins) are hydrocarbons — compounds that contain hydrogen and carbon *only*.
- Their molecules are saturated — they contain only single bonds
- They form a homologous series, with the general formula C_nH_{2n+2}

The first four alkanes in the series have names that are unrelated to the number of carbon atoms:
- methane, CH_4
- ethane, CH_3CH_3
- propane, $CH_3CH_2CH_3$
- butane, $CH_3CH_2CH_2CH_3$

For alkanes with five or more carbon atoms, the names are related to the number of carbon atoms in the chain: pentane, hexane, heptane, octane etc.

Structural isomerism

Structural isomers are compounds that have the same molecular formula but different structural formulae. The first three alkanes, CH_4, CH_3CH_3 and $CH_3CH_2CH_3$, have only one possible structure in each case. However the fourth, $CH_3CH_2CH_2CH_3$, has a structural isomer in which the atoms are arranged differently:

$$CH_3CH_2CH_2CH_3 \qquad \qquad CH_3CHCH_3$$

Butane

$$\underset{\displaystyle CH_3}{|}$$

2-methylpropane

In the case of the alkanes, the structural isomers are also alkanes.

However, structural isomers may be of different types. Thus, the molecular formula C_2H_6O could represent ethanol, CH_3CH_2OH, or the totally different methoxymethane, CH_3OCH_3, which is an ether. This is sometimes called **functional group isomerism**.

The number of structural isomers rises rapidly with the number of carbon atoms in the compound. For example, pentane has two structural isomers, so there are three structures that have the molecular formula C_5H_{12}.

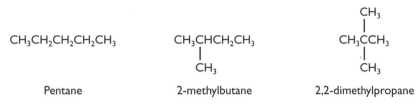

The number of structural isomers rises rapidly with the number of carbon atoms in the compound, as illustrated in the table below:

Alkane	Number of isomers
C_8H_{18}	18
$C_{10}H_{22}$	75
$C_{12}H_{26}$	355
$C_{14}H_{30}$	1858
$C_{20}H_{42}$	366 319
$C_{25}H_{52}$	36 797 588
$C_{30}H_{62}$	4 111 846 763
$C_{40}H_{82}$	62 491 178 805 831

Tips Because the number of potential structures can be large (even for small compounds), you should always use the structural formula to eliminate ambiguity.

Fuels

Fuels are obtained from the **fractional distillation** of crude oil. This process usually produces too little of the petrol and diesel fractions and too much of the heavier fuel and lubricating oil fractions. The latter are passed over suitable catalysts, with heating, in order to crack (or break) the molecules into smaller alkanes. Cracking also produces important alkenes used for chemical synthesis, which are too valuable to burn.

Liquid and gaseous fuels

The advantages and disadvantages of liquid and gaseous fuels depend on:
- whether the fuel has to be carried with the device using it (as in motor vehicles) or whether it can be piped (as in domestic gas supplies)
- any safety aspects peculiar to the particular fuel, especially the handling of potentially explosive gases (these do not include flammability, which some exam candidates seem to think is a problem — non-flammable fuels are rare)
- the energy yield per unit volume or unit mass (shown in the table below)

AS Chemistry

Fuel	M_r	$\Delta H_c/\text{kJ mol}^{-1}$	$\Delta H_c/\text{kJ cm}^{-3}$	$\Delta H_c/\text{kJ g}^{-1}$
Hydrogen(g)	2	−286	−0.012	−143
Methane(g)	16	−890	−0.037	−55.6
Butane (g)	58	−2877	−0.12	−49.6
Butane(l)	58	−2877	−29.8	−49.6
Octane(l)	114	−5470	−33.8	−48.0
Ethanol(l)	46	−1367	−23.4	−29.7

(The molar volume of the gases is taken as 24 dm³.)

Hydrogen and methane
Hydrogen and methane have a high energy yield per gram, but are not dense and therefore large volumes of gas are needed. They are used where the fuel can be piped to the point of use.

Hydrogen can be used in cars — it is compressed and adsorbed in a suitable metal sponge. However, the range of such vehicles is limited and the relatively heavy gas canisters take up a lot of space. There are presently few hydrogen filling stations.

All gaseous fuels present special handling requirements to avoid leakage of flammable or explosive gases. However, these problems have been solved.

Environmental factors Hydrogen burns to give water only. Remember, though, that the means of producing hydrogen (electrolysis of brine or reaction of methane with steam) consumes energy, so the environmental advantages of burning hydrogen have to be balanced by the various costs of its production. Methane is used in fixed systems and is burned efficiently, so that little carbon monoxide pollution results.

Butane
Butane has a high energy yield per gram, but the gas is not dense. It is easily lique-fied by compression to give liquefied petroleum gas (LPG). Butane can be used in cars that are adapted for it — the number of filling stations selling it is rising. It is also used in tanks and cylinders as a constituent of Calor gas. The liquid form vaporises readily and requires special handling facilities.

Octane
Octane has a high energy yield per gram and is the densest of the fuels listed. Therefore, manageable volumes of the fuel can be carried in motor vehicles. It is universally available. The liquid is volatile, but not so volatile that special handling is needed.

Environmental factors The yield of carbon monoxide from internal combustion engines is high. Most of this can be removed by the use of catalytic converters, in which carbon monoxide reacts with the nitrogen oxides that are also produced, to give nitrogen and carbon dioxide.

Ethanol

Ethanol has a high energy yield per gram, but it is not as dense as octane. Therefore, larger volumes of liquid are needed compared with octane. Ethanol is made from oil, if that is available, so it is not a cost-effective fuel for countries that have relatively cheap oil. Countries such as Brazil that have no oil but have a lot of sugar cane can make cheap, fuel-grade ethanol by fermentation. Car fuel in Brazil contains 20% ethanol and 80% ordinary petrol.

Environmental factors Ethanol is a clean fuel, producing little carbon monoxide.

Reactions of alkanes

Reaction with oxygen

In a plentiful supply of oxygen or air, alkanes burn on ignition to give carbon dioxide and water. The reactions are strongly exothermic, hence the use of alkanes as fuels:

* Methane is natural gas:

$$CH_4 + 2O_2 \rightarrow CO_2 + 2H_2O + heat$$

* Petrol is a mixture that can be represented by octane:

$$C_8H_{18} + 12\tfrac{1}{2}O_2 \rightarrow 8CO_2 + 9H_2O + heat$$

The main constituent in petrol is actually 2,2,4-trimethylpentane, which is an isomer of octane. In internal combustion engines, *complete* combustion does not occur; carbon and carbon monoxide are produced as well as carbon dioxide. In limited oxygen, all of the hydrogen in a given molecule is always oxidised — hydrogen is *never* a product.

Reaction with chlorine or bromine

In the presence of ultraviolet (UV) light, alkanes react with chlorine or bromine to give halogenoalkane. In strong UV or focused sunlight the reaction is explosive.

The reaction involves **radical substitution**. There is a mixture of products, because radicals (in this case halogen atoms) are reactive and reactions with the first halogenoalkane produced occur, resulting in further substitution. Methane reacting with chlorine gives CH_3Cl initially, which reacts to give, successively, CH_2Cl_2, $CHCl_3$ and CCl_4. Direct halogenation is not a good way of making single compounds, but is useful for making solvents, which can be mixtures.

$$CH_4 + Cl_2 \rightarrow CH_3Cl + HCl$$

$$CH_4 + Br_2 \rightarrow CH_3Br + HBr$$

Free radical substitution

The **mechanism** of a reaction is a way of representing how electrons are moved from the reactants to form the products. There are seven types of mechanism studied at AS and A2. The halogenation of alkanes involves **free radical substitution**; it is also a **chain reaction**.

Curly arrows are used to show the movement of electrons. They must be drawn with precision. The tail of the arrow must come from the electron(s); the head must be where the electrons go, to form the bond. The movement of a single electron is shown by a half-headed arrow; that of an electron pair by a full-headed arrow.

A **radical** is an atom or group of atoms with an unpaired electron. Most radicals are very reactive.

The first step in the chlorination of methane produces chlorine radicals (atoms), the Cl–Cl bond having been broken by UV light. The single electron is shown as a dot. This type of bond breaking, in which one electron from the bond goes to each of the fragments produced, is called **homolytic fission**.

The first **initiation step** produces the chlorine radicals:

$$Cl \longrightarrow Cl \qquad \longrightarrow \qquad 2Cl\bullet$$

The feature of a chain reaction is that the **propagation steps** that follow initiation produce one radical for every radical consumed. A chlorine radical attacks the methane molecule, producing a methyl radical and hydrogen chloride:

$$H_3C \longrightarrow H \quad \bullet Cl \qquad \longrightarrow \qquad \bullet CH_3 + HCl$$

The methyl radical then attacks a chlorine molecule, to give chloromethane and a chlorine radical:

$$Cl \longrightarrow Cl \quad \bullet CH_3 \qquad \longrightarrow \qquad \bullet Cl + ClCH_3$$

The chlorine radical attacks a methane molecule, and so the chain is propagated.

There are several possible **termination steps** where two radicals react to form a molecule, but not further radicals, thus ending the chain propagation:

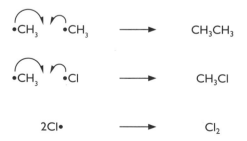

$$\bullet CH_3 \quad \bullet CH_3 \qquad \longrightarrow \qquad CH_3CH_3$$

$$\bullet CH_3 \quad \bullet Cl \qquad \longrightarrow \qquad CH_3Cl$$

$$2Cl\bullet \qquad \longrightarrow \qquad Cl_2$$

The mechanism for the bromination of methane is similar.

Alkenes

Alkenes are a homologous series with the general formula C_nH_{2n}. An alkene also contains a C=C double bond. Because of this double bond, alkenes are **unsaturated** compounds.

The double bond consists of two bonds that are not identical. Head-on overlap of carbon orbitals gives a σ-bond, in which the electron density is coaxial with the C–C internuclear axis. There is also sideways overlap between *p*-orbitals on the carbon atoms, which gives a π-bond as the other part of the double bond. The σ-bond is shorter than in saturated compounds (e.g. ethane), so is somewhat stronger; the σ-bond component accounts for about two-thirds of the overall strength of the C=C bond.

Geometric isomerism

Geometric isomerism results from restricted rotation about a carbon–carbon double bond, provided that the groups on a given carbon in the C=C bond are not the same.

Thus in the diagram above 'a' and 'e' must be different, as must 'b' and 'd'. The groups do not *all* have to be different from one another. The diagram below shows two geometric isomers:

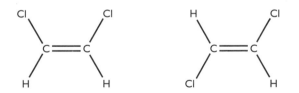

These are *cis-* and *trans-*1,2-dichloroethene. In *cis-*isomers the substituent groups, in this case the same, are on the same side of the C=C bond.

Sideways overlap of the *p*-orbitals to give a π-bond means that there is no rotation about C=C bonds, except at high temperatures.

The E–Z system of naming geometric isomers

If there are only two substituents (other than hydrogen) on the C=C bond, the *cis-trans* naming system works well. In the case of compounds that have more than two different groups around the C=C bond — for example, 2-bromobut-2-ene, the two forms of which are shown below — it is better to name them using a different system.

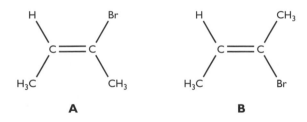

A   **B**

The rules are based on those of Cahn, Ingold and Prelog for naming chiral compounds. The substituent groups are ordered in a priority determined by the atomic mass of the atom directly bonded to a carbon atom. Taking isomer **A** above, hydrogen has atomic mass 1.0 and carbon has atomic mass 12.0. Thus, the carbon in the methyl group has the higher priority on the left-hand carbon. On the right-hand carbon, bromine (atomic mass 79.9) has higher priority than carbon (atomic mass 12.0). The two high-priority substituents are, therefore, the methyl (CH$_3$) group and the bromine, which are on opposite sides of the C=C bond. This isomer is the *entgegen* isomer, or *E*-isomer.

In isomer **B**, the high-priority groups are on the same side of the C=C bond, therefore, this is the *zusammen* or *Z*-isomer.

In fact, 2-bromobut-2-ene could be named using the *cis–trans* nomenclature; isomer **A,** with the methyl groups on the same side, being the *cis*-isomer. If one of these methyl groups were changed to an ethyl group then it would not be possible to assign *cis*- or *trans*- to the isomers. If all four groups are different as in F and G below:

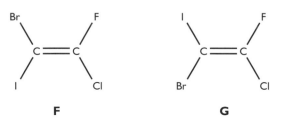

F   **G**

then only the *E–Z* system will work. You should convince yourself that **F** is the *Z*-isomer.

If the groups get larger (e.g. ethyl or propyl), the rules may need to be extended to adjacent atoms.

Addition reactions
Propene CH$_3$CH=CH$_2$ is a good exemple since it is an unsymmetrical alkene, i.e. the substituents on each end of the C=C bond are not the same. This is significant in some reactions.

Reaction with hydrogen

Propene reacts with hydrogen at room temperature in the presence of a platinum catalyst, or at 150°C with a nickel catalyst, to give propane:

$$CH_3CH=CH_2 + H_2 \rightarrow CH_3CH_2CH_3$$

This particular reaction is not useful industrially — alkenes are far more valuable than alkanes, since they are more reactive. However, it is the basis of margarine manufacture, and reduction of C=C bonds in more complex molecules is often required.

Reaction with halogens

Propene reacts with chlorine or bromine in the gas phase, or in an inert solvent (e.g. CCl_4), at room temperature:

$$CH_3CH=CH_2 + Br_2 \rightarrow CH_3CHBrCH_2Br$$

The product is 1,2-dibromopropane. The brown bromine is decolourised.

The reaction of bromine water (aqueous bromine) with alkenes is used to test for the presence of a C=C bond — the orange bromine water is decolourised. BrOH is added, not bromine itself:

$$CH_3CH=CH_2 + BrOH \rightarrow CH_3(Br)CHCH_2OH$$

Reaction with hydrogen halides

Propene reacts with hydrogen bromide in the gas phase, or in an inert solvent, at room temperature to form 2-bromopropane (see below) as the major product, with some 1-bromopropane.

$$CH_3CH=CH_2 + HBr \rightarrow CH_3CHBrCH_3$$

The mechanism is **electrophilic addition**.

The hydrogen goes to the carbon atom in the double bond that already has the most hydrogen atoms on it. This is Markovnikoff's rule; this 'rule' is *not* an explanation and there are exceptions to it. It is, however, a useful guide. The reasons for the orientation of the addition are dealt with below, when considering the mechanism.

Hydrogen chloride and hydrogen iodide react with propene in a similar manner to hydrogen bromide.

Reaction with potassium manganate(vii) solution

Propene reacts with potassium manganate(vii) solution, whether acidic (in H_2SO_4), neutral or alkaline (in Na_2CO_3), to give propane-1,2-diol:

$$CH_3CH=CH_2 \xrightarrow{\text{KMnO}_4} CH_3CH(OH)CH_2OH$$

The precise nature of this reaction is still not known, so a balanced equation is not written. The purple $KMnO_4$ solution is usually converted to a brown sludge of MnO_2.

Mechanisms
Electrophilic addition of bromine to ethene

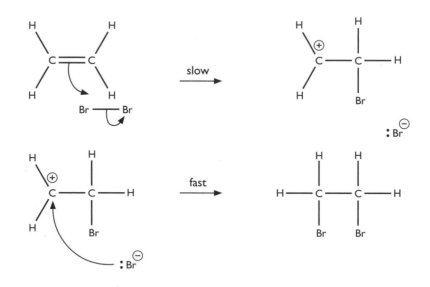

Electrophilic addition of hydrogen bromide to ethene

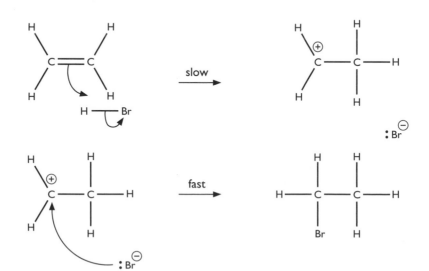

content guidance

Electrophilic addition of hydrogen bromide to propene

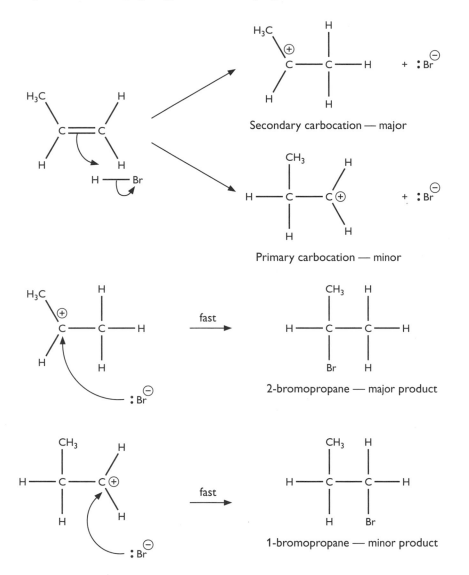

Secondary carbocation — major

Primary carbocation — minor

2-bromopropane — major product

1-bromopropane — minor product

Polymerisation of alkenes

Polyalkenes are made by radical addition reactions. A radical initiator such as a peroxide or oxygen is heated with the alkene. The first step is dissociation of the initiator into radicals:

$$R - R \longrightarrow 2R\bullet$$

The radical then attacks the alkene to form a new radical, and so on:

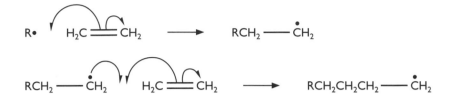

Because radicals are very reactive, they can attack growing chains at any point, forming branches and cross-links between the chains. The reaction stops when two radicals combine. Polymers contain molecules with a wide variety of chain lengths, so they soften over a range of temperatures rather than melting sharply.

Alkene monomers can contain substituent groups other than hydrogen. These groups *never* form part of the central carbon chain. This is illustrated in the table below:

Monomer	Polymer	Name
$CH_2\!=\!CH_2$	$-\!(CH_2\!-\!CH_2)_n\!-$	Poly(ethene)
$CH_2\!=\!CHCH_2$	$-\!(CH_2\!-\!\underset{\mid}{\underset{CH_3}{CH}})_n\!-$	Poly(propene)
$CH_2\!=\!CHCl$	$-\!(CH_2\!-\!\underset{\mid}{\underset{Cl}{CH}})_n\!-$	Poly(chloroethene), PVC
$CF_2\!=\!CF_2$	$-\!(CF_2\!-\!CF_2)_n\!-$	Poly(tetrafluoroethene), PTFE

Polymers and resources

Polymers are made from oil and need energy for their manufacture. They are unreactive. There are many different varieties that are not always easily distinguishable. Burning polymers can gives rise to toxic fumes. With the exception of PTFE, polymers are cheap.

Wrapping food in polymers has improved its keeping quality and has, therefore, reduced food waste and the incidence of food-related illness. In some cases, paper products could be used instead of polymers — unlike oil, paper is a renewable resource. However, the energy consumption in paper manufacture is significant. It also requires considerable quantities of water, the supply of which can be more of a problem than finding suitable trees — most of these are grown as a crop. Paper is never made from tropical hardwoods, so it does not contribute to rainforest decline; it is made from the timber of conifers. The soil in which conifers are grown emits nitrogen oxides into the atmosphere and can contribute a greater volume of nitrogen oxides to the locality than motor vehicles do. So, paper cups are not necessarily better than poly(styrene) ones.

All these points have to be borne in mind when using polymer products or considering alternatives. Polymers can all be recycled, but recycling has energy costs in terms of transport and reprocessing of a low-value material. Sorting the different polymer varieties is expensive. On the other hand, polymers are durable and so take up landfill space, are visually polluting and in the seas can cause seabirds and other aquatic creatures to die from choking if they swallow plastic waste.

Given information on a particular product, you should weigh up the advantages and disadvantages of using it. You should recognise that there are no cost-free solutions to any of our efforts to make processes more sustainable.

Questions & Answers

This section contains multiple-choice and structured questions similar to those you can expect to find in Unit Test 1. The questions given here are not balanced in terms of types of question or level of demand — that is, they are not intended to typify real papers, only the sorts of questions that could be asked.

In the examinations, the answers are written on the question paper. Here, the questions are not shown in examination paper format.

The answers given are those that examiners would expect from a grade-A candidate. They are not 'model answers' to be regurgitated without understanding. In answers that require more extended writing, it is usually the ideas that count rather than the form of words used; the principle is that correct and relevant chemistry scores.

Before you start to read and use this section, re-read the material on answering examination questions in the introduction to this guide.

Examiner's comments

Responses to questions may have an examiner's comment, preceded by the icon ℮. The comments may explain the correct answer, point out common errors made by candidates who produce work of C-grade or lower, or contain additional material that could be useful to you.

Multiple-choice questions

See pages 10–12 of the introduction for details of how this type of question will appear on the papers.

Each of the questions or incomplete statements is followed by four suggested answers, A, B, C or D. Select the *best* answer in each case. The answers are given, with some commentary, after question 10.

1 The definition of the mole is:

A   the number of atoms in exactly 12 grams of the isotope $^{12}_{6}C$

B   the number of molecules in 24 dm^3 of a gas at 273 K and 1 atm

C   the number of molecules in 24 dm^3 of a gas at room temperature and 1 atm

D   the amount of any substance containing the same number of elementary entities as there are atoms in exactly 12 grams of the isotope $^{12}_{6}C$

2 Which one of the following equations is associated with the definition of the standard enthalpy of formation of carbon monoxide?

A $C(s) + \frac{1}{2}O_2(g) \rightarrow CO(g)$

B $C(graphite) + \frac{1}{2}O_2(g) \rightarrow CO(g)$

C $C(s) + O(g) \rightarrow CO(g)$

D $C(g) + \frac{1}{2}O_2(g) \rightarrow CO(g)$

3 If the average C–H bond enthalpy is $+x$, which one of the following represents a process with enthalpy change $+4x$?

A $C(g) + 4H(g) \rightarrow CH_4(g)$

B $CH_4(g) \rightarrow C(g) + 4H(g)$

C $CH_4(g) \rightarrow C(g) + 2H_2(g)$

D $C(s) + 2H_2(g) \rightarrow CH_4(g)$

4 The elements of group 1 (Li–Cs) are very electropositive because:

A   they all have small atomic radii

B   their ions all have low hydration enthalpies

C   they all have low first ionisation energies

D   they all have high second ionisation energies

5   The structure of butan-2-ol is represented by the skeletal formula:

A

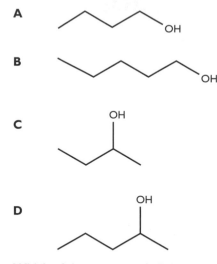

B

C

D

6   Which of the processes **A–D** is a propagation step in the reaction between chlorine and methane in the presence of ultraviolet light?

A   •CH_3 + Cl• → CH_3Cl

B CH_4 + Cl• → CH_3Cl + H•

C CH_4 + Cl_2 → CH_3Cl + HCl

D CH_4 + Cl• → •CH_3 + HCl

7   Ethoxyethane, usually known as ether, is often used to extract organic compounds from aqueous solution. Which of its properties **A–D** does *not* favour its selection for this use?

A   it dissolves most organic compounds easily

B   it is immiscible with water

C   its air/vapour mixture is explosive

D   it is inert to most types of reaction

8   In an experiment to measure the enthalpy of neutralisation of hydrochloric acid, 25.0 cm³ of hydrochloric acid containing 0.050 mol of HCl was reacted with 25.0 cm³ of sodium hydroxide solution of the same concentration in an expanded poly-styrene cup. The temperature rise was 13.7 °C. The heat capacity of the solution is 4.2 J g^{-1}°C^{-1} and the density is 1.00 g cm^{-3}. What is the enthalpy of neutralisation of hydrochloric acid (to 3 s.f.)?

A   −57.5 kJ mol^{-1}

B   +57.5 kJ mol^{-1}

C   −28.8 kJ mol^{-1}

D   −2.88 kJ mol^{-1}

9   The use of large amounts of disposable poly(alkene) packaging has been criticised because:

   A   a large amount of fossil fuel is burnt during its manufacture

   B   the combustion of poly(alkenes) produces dangerous fumes

   C   poly(alkenes) degrade to produce toxic materials

   D   a large amount of fossil fuel is converted to the necessary monomers, such as ethene

10 The set of ions in which the members all have the same electron configuration is:

   A Fe^{2+}, Fe^{3+}

   B N^{3-}, O^{2-}, F^-

   C $SO_4^{2-}, SeO_4^{2-}, TeO_4^{2-}$

   D F^-, Cl^-, Br^-

Answers

1   D

   The general definition of the mole has to be able to cope with particles other than atoms. Therefore the definition is **D**, with the proviso that 'the elementary entities must be specified and may be atoms, molecules, ions, electrons, other particles, or specified groups of such particles' (International Committee for Weights and Measures, 1971 and 1980). Option **A** is a statement related to the value of the Avogadro constant.

2   B

   This answer makes clear that the most stable form of the element carbon under standard conditions is graphite. This makes option **B** of higher quality than **A**, which is nearly right. Hence the exhortation in the introduction to read the questions carefully and not to make up your mind too soon.

3   B

   The bond enthalpy relates to the energy change in the production of atoms, so isolated atoms must be the product if all four of the bonds in methane are broken.

4   C

   The low ionisation energies mean that this endothermic change can be recovered from the bonding produced when the cation and the anion attract one another. Options **A** and **B** are both the opposite of the truth, but **D** is true. Some of the incorrect answers to the question set can be statements which in themselves are true. This is a common source of error in multiple-choice papers.

5 C

📖 The chain contains four carbon atoms with the −OH group on the second carbon. Formula A is butan-1-ol, and B and D are isomers of the 5-carbon alcohol, pentanol.

6 D

📖 The propagation steps must produce a radical for every radical that is used, so, on this basis alone, **B** could be an answer. However, answers do have to correspond with what actually happens in the universe.

7 C

📖 This is a *hazard* that must be overcome if ether is to be used. The *risk* from its explosive nature depends on how the ether is used. Using a fume cupboard and ensuring that there are no naked flames are obvious solutions. Perhaps there are other less hazardous solvents that, in a given circumstance, could be used instead.

8 A

📖 The calculation is:

$$\Delta H = \frac{-50.0\,g \times 4.2\,Jg^{-1}{}^{\circ}C^{-1} \times 13.7^{\circ}C}{0.050\,mol}$$

$= -57.540\,J\,mol^{-1}$ or $-57.5\,kJ\,mol^{-1}$ to the required 3 s.f.

The other values are distracters arising from common errors in this type of calculation. The value is negative because the temperature has risen and so the reaction must be exothermic.

9 D

📖 None of answers A–C is true.

10 B

📖 All of these ions have the electron configuration $1s^2\,2s^2\,2p^6$, i.e. that of neon.

Structured questions

Question 11

Crude oil contains a large proportion of alkanes, which form a homologous series.

(a) (i) Explain the meaning of the term 'homologous series', using the alkanes as your example. (2 marks)

(ii) Alkanes exhibit structural isomerism. Draw the structural formulae of the three structural isomers of the alkane C_5H_{12}. (3 marks)

(b) Methane reacts with chlorine in the presence of UV light to give chloromethane and hydrogen chloride:

$$CH_4 + Cl_2 \rightarrow CH_3Cl + HCl$$

(i) What type of reaction is this? (1 mark)

(ii) The first step of the mechanism involves homolytic fission. Explain the meaning of this term. (1 mark)

(iii) Give the mechanism for the reaction between methane and chlorine, making clear which are the initiation, propagation and termination steps. (6 marks)

Total: 13 marks

Answer to question 11

(a) (i) A series of compounds with the same general formula, in this case C_nH_{2n+2} ✓, and showing similar chemical properties ✓.

🖉 The inclusion of the particular general formula is necessary to make the answer refer to the alkanes, as required by the question. If asked about a specific type of compound, make sure that your answer is specific, not general.

(a) (ii)

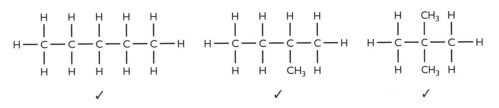

(b) (i) (Free) radical substitution ✓

(b) (ii) Breaking a covalent bond so that one electron goes to each of the two species formed, which are radicals ✓

(b) (iii) Initiation:

 $\longrightarrow$ $2Cl\bullet$ ✓

Propagation:

$CH_4 + Cl\bullet \rightarrow \bullet CH_3 + HCl$ ✓

$\bullet CH_3 + Cl_2 \rightarrow CH_3Cl + Cl\bullet$ ✓

Termination:

$2Cl\bullet \rightarrow Cl_2$ ✓

$Cl\bullet + \bullet CH_3 \rightarrow CH_3Cl$ ✓

$2 \bullet CH_3 \rightarrow C_2H_6$ ✓

Question 12

(a) State Hess's law.   (2 marks)

(b) Define the term 'standard enthalpy of combustion', making clear the meaning of 'standard' in this context.   (3 marks)

(c) The equation for the combustion of ethanol in air is:

$$C_2H_5OH(l) + 3O_2(g) \rightarrow 2CO_2(g) + 3H_2O(l)$$

Calculate the enthalpy change for this reaction using the average bond enthalpy values given in the table.   (3 marks)

Bond	Average bond enthalpy/kJ mol^{-1}	Bond	Average bond enthalpy/kJ mol^{-1}
C–H	+412	C–C	+348
C–O	+360	O–H	+463
O=O	+496	C=O	+743

Total: 8 marks

Answer to question 12

(a) The heat energy/enthalpy change in a chemical reaction is independent of the route used to go from the reagents to the products ✓ provided that the initial and final states are the same ✓.

🖉 The term 'energy change' alone should not be used — this refers to a reaction at constant volume, not constant pressure.

(b) The heat energy change per mole ✓ for the complete combustion of a substance in excess oxygen ✓ at 1 atm pressure and stated temperature ✓.

💬 In practice, quoting the temperature as 298 K would also gain credit, although this particular temperature is not part of the definition of the standard state. *Heat* energy change, not energy change alone, is important. The energy change for a reaction is given the symbol ΔU and is measured at constant volume, not constant pressure.

(c) There is no standard notation for the average bond enthalpy. Using $D(X–X)$ to present it for the X–X bond:

ΔH = (sum of bond energies of reagents) – (sum of bond energies of the products) ✓

$= [5D(C–H) + D(C–O) + D(C–C) + D(O–H) + 3D(O=O)] – [4D(C=O) + 6D(O–H)]$ ✓

$= [(5 \times 412) + 360 + 348 + 463 + (3 \times 496)] – [(4 \times 743) + (6 \times 463)] = -1031\,\text{kJ mol}^{-1}$ ✓

💬 Note that brackets are used separate the enthalpy values for each type of bond. The absence of such working is characteristic of C-grade answers and is a high-risk strategy. If the answer only is shown and it is wrong, you cannot get marks for intermediate steps.

It is also important to be able to see the chemical background to the calculation. Examiners are better disposed towards candidates whose work is transparent.

Be careful to choose the correct values — it is common to find, for example, that the C–O bond strength is used for the C=O bond.

Question 13

(a) (i) Complete the electron configuration of a sulfur atom: $1s^2$.... (1 mark)
(ii) State the number of neutrons in the nucleus of an atom of ^{32}S. (1 mark)
(b) (i) Define the term 'first electron affinity'. (2 marks)
(ii) The following equation represents the change occurring when the second electron affinity of sulfur is measured:
$S^-(g) + e^- \rightarrow S^{2-}(g)$

Explain why the second electron affinity of an element is always endothermic.
(2 marks)

Total: 6 marks

Answer to question 13

(a) (i) $1s^2\,2s^2\,2p^6\,3s^2\,3p^4$ ✓

(a) (ii) $32 - 16 = 16$ ✓

🅔 This question requires you to use the periodic table that is provided with all question papers. It is, of course, another way of finding out if you know what the terms atomic number and mass number mean.

(b) (i) The energy change for the addition of one electron to each atom in one mole of atoms ✓ in the gas phase ✓.

🅔 This could be answered by means of one statement with an equation, namely: the energy change per mole ✓ for the process:

$$X(g) + e^- \rightarrow X^-(g) \checkmark$$

(b) (ii) The negative electron is being added to a negative ion ✓ and the two negative charges repel ✓.

🅔 Succeeding electron affinities would also be positive (endothermic); in practice, simple ions of charge greater than 3– are not seen.

Question 14

(a) **When a sample of copper is analysed using a mass spectrometer, its atoms are ionised, accelerated, and then separated according to their mass/charge (m/e) ratio.**
   (i) **Explain how the atoms of the sample are ionised.** (2 marks)
   (ii) **State how the resulting ions are accelerated.** (1 mark)
   (iii) **State how the ions are separated according to their m/e values.** (1 mark)

(b) **For a particular sample of copper, two peaks were obtained in the mass spectrum, showing an abundance of 69.10% at m/e 63, and 30.90% at m/e 65.**
   (i) **Give the formula of the species responsible for the peak at m/e 65.** (1 mark)
   (ii) **State why two peaks, at m/e values of 63 and 65, were obtained in the mass spectrum.** (1 mark)
   (iii) **Calculate the relative atomic mass of this sample of copper to three significant figures, using the data given above.** (2 marks)

Total: 8 marks

Answer to question 14

(a) (i) Fast-moving *or* energetic electrons strike the atoms ✓ removing electrons from the sample atoms, forming positive ions ✓.

🅔 The copper sample would be on a heated probe; although the vapour pressure of copper is very small, the number of atoms volatilised at the extremely low pressure in the mass spectrometer is large enough to be detected.

(a) (ii) In an electric *or* electrostatic field ✓

(a) (iii) By a magnetic field ✓

(b) (i) $^{65}Cu^+$ ✓

🗩 Many candidates forget to put the essential positive charge when giving the formulae of the ions detected by a mass spectrometer.

(b) (ii) Because naturally-occurring copper contains two isotopes of mass 63 and 65 ✓

(b) (iii) Relative atomic mass = $\dfrac{(63 \times 69.10) + (65 \times 30.90)}{(69.10 + 30.90)}$ ✓ = 63.6 ✓

🗩 The question asks for *three* significant figures. If you give 63.618 or 63.62 you will not score the mark. This is a silly way to lose marks.

Question 15

(a) Explain the following observations. Include details of the *bonding* in, and the *structure* of, each substance.
   (i) The melting temperature of diamond is much higher than that of iodine. (5 marks)
   (ii) Sodium chloride has a high melting temperature (approximately 800°C) (3 marks)
(b) Explain in terms of its structure and bonding why aluminium is a good conductor of electricity. (3 marks)

Total: 11 marks

Answer to question 15

(a) (i) Iodine is molecular covalent ✓ so has weak intermolecular forces ✓; diamond is giant covalent ✓, so has strong intramolecular covalent bonds throughout the crystal lattice ✓. The stronger the forces, the more energy is needed to overcome them to melt the crystal ✓.

🗩 When discussing the bonding of covalent substances, it is important to distinguish between *inter*molecular and *intra*molecular forces. Covalent bonds are strong.

(a) (ii) Sodium chloride has a lattice ✓ of oppositely charged ions ✓ with strong attractions throughout the lattice ✓.

🗩 The extension of the attractions throughout the lattice is important. There are no pairs of ions that can be selected over any other pair; the attraction is uniform throughout the crystal. There are also repulsions in the lattice between ions of the same charge; the attraction is a net attraction.

(b) Metal ions are in a lattice ✓ bonded by attraction to delocalised electrons *or* embedded in a sea of electrons ✓. Mobile electrons enable conduction ✓.

📖 Make sure that you say that the electrons are mobile. Just saying that they form a 'sea' is not enough. It is also important to say that the electrons are attracted to the ions.

Question 16

(a) Lithium chloride, LiCl, can be made by the reaction of lithium with chlorine:

$2Li(s) + Cl_2(g) \rightarrow 2LiCl(s)$

    (i) Calculate the maximum mass of lithium chloride that can be made from 35 g of lithium.   (2 marks)

    (ii) Calculate the concentration in $mol\,dm^{-3}$ of the solution that would be obtained if this mass of lithium chloride were dissolved in water to make $5.00\,dm^3$ of solution.   (2 marks)

    (iii) Calculate the volume of chlorine gas required to react with 35 g of lithium. The molar volume of a gas at the temperature and pressure of the experiment is $24\,dm^3$.   (2 marks)

(b) Describe the structure of solid lithium metal and explain why it conducts electricity.   (3 marks)

(c) (i) Define the term 'first ionisation energy'.   (3 marks)

    (ii) Explain why the first ionisation energy of lithium is less endothermic than the first ionisation energy of neon.   (3 marks)

    (iii) Explain why the first ionisation energy of sodium is less endothermic than that of lithium.   (2 marks)

Total: 17 marks

Answer to question 16

(a) (i)   amount of lithium used $= 35\,g \div 7.0\,g\,mol^{-1} = 5.0\,mol$ ✓

           mass of LiCl produced $= 5.0\,mol \times 42.5\,g\,mol^{-1} = 212.5\,g$ ✓

(a) (ii)   concentration $= 5.0\,mol \div 5.00\,dm^3 = 1.0\,mol\,dm^{-3}$ ✓

(a) (iii) 1 mol Li reacts with $0.5\,mol\,Cl_2$ ✓

           $\therefore$ volume Cl_2 required $= 0.5 \times 5.0\,mol \times 24\,dm^3\,mol^{-1} = 60\,dm^3$ ✓

📖 Note that the use of units throughout the calculations enables you to check that what you are doing is correct, since the units of the answer will also be correct.

(b) The metal crystal has a lattice of ions ✓ that are attracted to a 'sea' of delocalised electrons ✓. These electrons can drift through the lattice and so enable the metal to conduct electricity ✓.

📖 The 'attracted to' point is important.

(c) (i) The energy change ✓ for the formation of a mole of gaseous unipositive ions ✓ from a mole of gaseous atoms ✓.

or

The energy change per mole ✓ for the process:

$E(g) \rightarrow E^+(g) + e^-$ ✓✓

🖉 If you can define things through equations, do so. It is often quicker and clearer. The equation scores 2 marks because it includes both the positive ion and the gas phase.

(c) (ii) The attraction for the outer electron is a combination of the nuclear charge, the repulsions from inner electrons, and the size of the atom ✓. The lithium atom is larger than the neon atom and the nuclear charge of lithium is lower than that of neon ✓. The lithium electron is, therefore, further from a less positive nucleus and so it is less attracted by it ✓.

(c) (iii) Sodium is a larger atom than lithium. Therefore, the greater nuclear charge is offset ✓ by the increased number of shells, so there are increased repulsions from the inner electrons ✓.

🖉 The idea of repulsion from inner electrons is sometimes expressed as shielding of the nuclear charge from the outer electrons by the inner ones. Some students find the notion of repulsions easier to visualise.

The term 'less endothermic' is preferable to 'lower than'. The positive and negative signs in thermochemistry are conventions that show the direction of movement of heat energy.

Question 17

(a) State Hess's law. (2 marks)

(b) Methane burns in oxygen:

$CH_4(g) + 2O_2(g) \rightarrow CO_2(g) + 2H_2O(g)$

    (i) Use your *Data Booklet* to find appropriate values of the bond enthalpies needed to calculate the enthalpy change for this reaction. (2 marks)

    (ii) Calculate this enthalpy change. (3 marks)

    (iii) If the enthalpy change for the reaction is calculated using standard enthalpies of formation its value is $-890\,kJ\,mol^{-1}$. What is meant by the term 'standard'? (2 marks)

    (iv) Explain why the value given in (iii) is different from that obtained in (ii).

(2 marks)

Total: 11 marks

Answer to question 17

(a) The enthalpy change for a reaction is independent of the path taken ✓ provided that the initial and final states are the same ✓.

(b) (i) Bond enthalpies/kJ mol^{-1}: C–H 435, O=O 498, C=O 805, O–H 464.

🕮 The *Data Booklet* is a new feature for Edexcel examinations and the ability to use it correctly is essential. In this question, the bond enthalpy values for the particular molecules involved are needed, so the ability to select these values, rather than average ones, is important.

(b) (ii) bonds broken = 4(C–H) + 2(O=O)

so, energy used = (4 × 435) + (2 × 498) = 2736 kJ mol^{-1} ✓

bonds made = 2(C=O) + 4(O–H)

so, energy gained = (2 × 805) + (4 × 464) = 3466 kJ mol^{-1} ✓

ΔH = 2736 – 3466 kJ mol^{-1} = –730 kJ mol^{-1} ✓

🕮 In every calculation that you do, you must include the units. It doesn't matter whether there are marks available or not; you are dealing with physical quantities, and physical quantities have units. The aim is to understand chemistry — a separate idea from merely gaining marks in an examination.

(b) (iii) The standard state refers to all substances being in their thermodynamically most stable state ✓ at 1 atm pressure and a stated temperature ✓, usually 298 K.

🕮 Strictly speaking, the temperature is not part of the definition. However, if no other temperature is stated, 298 K is taken as the default temperature. The first scoring point must include 'most stable state' to allow for substances — for example, carbon — that, at 298 K, have several different solid states of different stabilities.

(b) (iv) Enthalpies of formation are given for a specific substance ✓ whereas bond enthalpies are average values for a range of compounds ✓.

Question 18

(a) Methane reacts with chlorine in the presence of ultraviolet light.
  (i) **Give the equation for the reaction that produces chloromethane.** (1 mark)
  (ii) **Give the mechanism for this reaction, showing the initiation and propagation steps, and one possible termination step other than that which re-forms chlorine.** (4 marks)
(b) Ethene reacts with bromine, but light is not needed.
  (i) **Give the equation for the reaction and state what you would see.** (2 marks)
  (ii) **Give the mechanism for the reaction of an alkene with bromine.** (3 marks)

Total: 10 marks

Answer to question 18

(a) (i) $CH_4 + Cl_2 \rightarrow CH_3Cl + HCl$ ✓

(a) (ii)   Initiation:

→   2Cl•

Propagation:

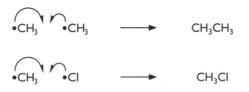

→   •CH$_3$ + HCl

→   •Cl + ClCH$_3$

Termination:

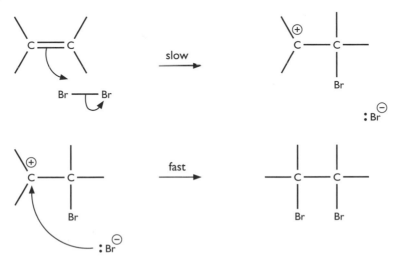

•CH$_3$ •CH$_3$ →   CH$_3$CH$_3$

•CH$_3$ •Cl   →   CH$_3$Cl

⌨ There is 1 mark for either of the *two* termination steps shown. The use of the 'half-headed' arrows is not strictly necessary, unlike in non-radical mechanisms where the use of the full-headed arrow *is* necessary.

(b) (i) $H_2C=CH_2 + Br_2 \rightarrow BrCH_2CH_2Br$ ✓

The brown colour of bromine is lost ✓.

(b) (ii)

slow

fast

⌨ There is 1 mark for the first two arrows, 1 mark for the intermediate carbocation and 1 mark for the attack of the bromide ion on the carbocation. Note that arrows are not decorative; they are there to show the movement of electron pairs, so they must start and finish in the proper places.